190 C

BARTÓK IN BRITAIN

BARTÓK IN BRITAIN

A Guided Tour

MALCOLM GILLIES

CLARENDON PRESS · OXFORD
1989

Oxford University Press, Walton Street, Oxford OX2 6DP
Oxford New York Toronto
Delhi Bombay Calcutta Madras Karachi
Petaling Jaya Singapore Hong Kong Tokyo
Nairobi Dar es Salaam Cape Town
Melbourne Auckland
and associated companies in
Berlin Ibadan

Oxford is a trade mark of Oxford University Press

Published in the United States
by Oxford University Press, New York

British Library Cataloguing in Publication Data
Gillies, Malcolm
Bartók in Britain: a guided tour.
1. Hungarian music. Bartók, Béla, 1881–
1945. Biographies
I. Title
780'.92'4
ISBN 0–19–315262–2

Library of Congress Cataloging in Publication Data
Gillies, Malcolm.
Bartók in Britain: a guided tour/Malcolm Gillies.
p. cm. Bibliography: p.
Includes index.
1. Bartók, Béla. 1881–1945. 2. Music—Great Britain—20th
century—History and criticism. I. Title.
ML410.B26G46   1988
780'.92'4—dc 19
[b]
ISBN 0–19–315262–2

Typeset by Cambrian Typesetters, Frimley, Surrey
Printed in Great Britain
at the University Printing House, Oxford
by David Stanford
Printer to the University

PROLOGUE

The characters of some musicians lie patent to the world. Liszt and Wagner are good examples. A century after their deaths we are still in danger of being overwhelmed by their continual, compulsive self-advertisement through every available medium. At the other pole are the introverted, 'mysterious' musicians who develop a reflective façade to shield their sensitive personalities from the harshness of the real world and only feel at ease behind its security. The Hungarian composer Béla Bartók was one such sensitive figure. As a colleague once commented, he presented to the world more the image of a dutiful bank clerk than that of one of the century's most creative musicians.

Bartók habitually stood apart from his surroundings, thereby giving an impression of arrogance to strangers, and of being the classic 'loner' to acquaintances. To him the rituals of self-promotion often associated with the 'great musician' were anathema, and journalistic demands for an explanation of the secrets of his art constituted a flagrant threat to his personal integrity. Although he did unbend a little in the presence of family or close friends, it was as if this lowering of his guard was a necessary and inevitable concession to the practicalities of day-to-day existence rather than a natural and desired state of his being. Hence 'the Bartók mystery' and the reputation of his unknowability. After his death, the critic Cecil Gray could only comment: 'Béla Bartók, in short, was completely inhuman. He hardly existed as a personality, but his impersonality was tremendous—he was the living incarnation and embodiment of the spirit of music.'[1]

The difficulty involved in probing this introverted character has encouraged many writers to concentrate solely on his music: to search for the keys to his aural treasure chests. To tackle the man—bound up in the chains of an exotic language and culture, as well as in his own personal coils—is an altogether more slippery proposition. As a result, our knowledge of Bartók and his life has not kept pace with the findings about his music. For English speakers, in particular, the Bartók presented in books, programme-notes, videos, and exhibitions is still essentially that Bartók unfolded by Halsey Stevens in his pioneering book of 1953.[2] Little has been done since to revise, elaborate, or correct that solid but dated interpretation.

Bartók in Britain attempts a fresh look at Bartók, not in a generalized, full-frontal assault upon his life, but in a 'depth study' into one of its more representative, geographically determined corners. There we are more likely

---

[1] *Musical Chairs* (London, 1948), 181.
[2] *The Life and Music of Béla Bartók* (New York, 1953; 2nd edn. New York, 1964).

to come across the fingerprints of the man, and have time to delve into the detail of his reactions and the reactions of others to him. On this 'guided tour' we follow Bartók methodically through nearly twenty concert tours of Britain which he undertook between 1904 and 1938. From youth to late middle age he guides us—from Bournemouth to inner Glasgow, from a girls' school on the Suffolk coast to the University College in Aberystwyth. We follow in detail his tussles with the London critics over the years as the fortunes of his music and his pianism fluctuate. His none-too-tactful battles with BBC bureaucracy are investigated, as are some more personal issues of his relationships with his British hosts.

On such a tour we must perforce learn much about Britain's musical life at the time. Bartók's visits naturally lead us from Edwardian times, when concert-going was still very much the preserve of upper sections of society, through the dramatic changes wrought by war, and the advent of radio and sounded films, to the era of greater professionalism and a more democratic listening public in the 1930s. Important changes also occurred in venues, programmes, and critical attitudes to music, and these need to be followed if the significance of Bartók's activities in the country is really to be grasped. Bartók and his music, therefore, assume something of the role of a constant in this study, behind which the relevant variables of Britain's musical life are depicted.

Beyond Bartók and Britain, however, this book has no sustained themes. The tour is packaged—the facts of history have seen to that—but not for any purpose of academic neatness or moral injunction. This may prove an irritation to those who like a good plot. Certainly, comparisons are drawn, characteristics established, and tendencies suggested, but the illogicalities, anomalies, chance events, and surprises found in real life are the very materials by which it is often possible to penetrate through Bartók's tough exterior to the private man and musician. As several Bartók studies have already shown, the simplicity of a few relentlessly pursued themes can well lead to the replacement of Bartók's façade with one manufactured by the author, or even the State. Both prevent us from progressing beyond a superficial understanding of this great musician and the life which he lived. As appropriate to an exploratory 'depth study', *Bartók in Britain* is concerned to expose the widest possible range of facts and the contexts necessary for their understanding: the diversity and detail of what was going on or being thought most of the time, within the deliberately narrow limits of the study. Through this intensity of focus we have an opportunity to glimpse something of the real Bartók at various stages of his life, and to appreciate the unpredictable ebb and flow of his fortunes in British musical circles.

The book's two parts have different purposes. Part I, Chapters 1 to 6, is arranged chronologically. It provides an account of Bartók's various visits and of British reactions to his music over the corresponding years. In Chapters 5 and 6 there is a slight deviation from this chronological path, the former chapter dealing with all Bartók's BBC engagements, the latter outlining his other activities in Britain during these same years. Part II consists of two studies which exploit the 'depth' approach more fully. Chapter 7, in documenting Bartók's relationship with the English critics Philip Heseltine and Cecil Gray, provides a more detailed and selective account of British appreciations of Bartók's music. Chapter 8 delves into Bartók's personal relations with the Arányi sisters, and their relevance to his performing activities. The Epilogue surveys British responses to Bartók's music between his last visit in 1938 and his death in 1945, by which stage the foundations of his present-day high profile were already being laid.

ACKNOWLEDGEMENTS

To the following institutions and individuals who have greatly assisted my research and have granted publication permissions, I express my thanks: Béla Bartók, jun. (Budapest), Péter Bartók (Homosassa, Florida), the British Broadcasting Corporation Written Archives Centre (Reading), The British Library (London), the Budapest Bartók Archive, János Demény (Budapest), Louis Szathmáry (Chicago).

To those listed below I am endebted for their critiques of my text and useful suggestions: Naomi Cumming (Adelaide), Belle Gillies (Canberra), Adrienne Gombocz (Budapest), Helen Kasztelan (Melbourne), David Pear (Melbourne), László Somfai (Budapest), Camilla Webster (Cambridge), András Wilheim (Budapest). In translation work I acknowledge the assistance of Atilla Ágoston, Kornél Bárdos, Ágnes Forró, Adrienne Gombocz (all of Budapest), Erzsébet Burián, Tamás Burián, Kerry Murphy (all of Melbourne).

Others who have substantially assisted my research or granted publication permissions are: David Adams (London), Verily Anderson (Cromer), Amanda Arrowsmith (Ipswich), Christopher Bornet (London), Ian and Kathleen Bott (London), the Bournemouth Symphony Orchestra, János Breuer (Budapest), Michael Brimer (Melbourne), Phyllis H. Brodie (Cape Town), Clare Brown (Reading), Patrick Cadell (Edinburgh), Adrienne Fachiri Camilloni (Florence), David Clegg (Lichfield), Thomas Colchester (Aldeburgh), David Cummings (London), Sandrey Date (Bournemouth), Peter Dennison (Melbourne), Kay Dreyfus (Melbourne), Sybil Eaton (London), M. J. Evans (West Malvern), Howard Ferguson (Cambridge), András Fodor (Budapest), Lillias M. Forbes (St Andrews), Iris Gibb (Henfield), Hallé Concerts Society (Manchester), Margot Hare (Saxmundham), Margaret Jacoby (Battle), John Lade (London), Fiona Lello (Lakeside, Cape), the Library of Congress (Washington), Maresa Macleod (Florence), Jennifer Marshall (Sydney), the National Library of Australia (Canberra), Yehudi Menuhin (London), the National Library of Scotland (Edinburgh), the former New York Bartók Archive, the New York Public Library, J. M. Nixon (West Malvern), Ian Parrott (Aberystwyth), the Peter Warlock Society (London), Laurie Pettitt (London), the Pierpont Morgan Library (New York), James Rushworth (Liverpool), Lorna Sanders (Battle), Eleonore Schacht (Zell), J. M. Stewart (Glasgow), Benjamin Suchoff (Palm Beach), Tibor Tallián (Budapest), Ronald Taylor (St Julians, Kent), Graham Thorne (London), Stuart Thyne (Edinburgh), the University of London Library, Arnold Whittall (London). Every effort has been made to contact copyright holders. I apologize to anyone who may have been overlooked.

Several passages of this book have appeared in a different form in the following journals: *Music and Letters* (parts of Chapters 2 and 3); *Music and the Teacher* (parts of Chapter 4); *Music Review* (Chapter 7). I am grateful to these journals' editors for permissions to include those sections.

The following publishers have generously granted permissions to reprint passages from the listed books: Faber and Faber Ltd. (János Demény, ed., *Béla Bartók Letters*; Eric Fenby, *Delius as I knew him*); Victor Gollancz Ltd. (Sir Henry Wood, *My Life of Music*); Houghton Mifflin Company (Agatha Fassett, *Béla Bartók's Last Years: The Naked Face of Genius*); Unwin Hyman (Eric Fenby, *Delius as I knew him*).

To the following I am indebted for permissions to reproduce illustrations: Battle and District Historical Society (nos. 6, 7); Béla Bartók, jun. (nos. 4, 5, 14); Ferenc Bónis (nos. 11, 12); the British Broadcasting Corporation (cartoons on fos. 69, 73, 86); László Somfai (on behalf of the Budapest Bartók Archive) (nos. 1, 2, 8, 11, 12, 13).

To the staff of Oxford University Press, in particular to the Music Books Editor, Bruce Phillips, I express my thanks for expert advice and support over the years since the idea of this book was first discussed.

From the institutions listed below I acknowledge the generous financial support which has made possible the realization of this project: the Association of Commonwealth Universities (London), the Australian Academy of the Humanities (Canberra), the Hungarian Government, the University of Melbourne, and the Victorian College of the Arts (Melbourne).

CONTENTS

LIST OF PLATES

Plates are between pages 32 and 33

ABBREVIATIONS

BBCWAC 47796	British Broadcasting Corporation Written Archives Centre (Reading), Accession no. 47796, Béla Bartók.
BBcl	*Bartók Béla családi levelei* [Béla Bartók Family Letters], ed. Béla Bartók, jun. (Budapest, 1981). In original Hungarian.
BBE	*Béla Bartók Essays*, ed. Benjamin Suchoff (London, 1976).
BBL	*Béla Bartók Letters*, ed. János Demény (London, 1971).
*BBlev.*i–v	*Bartók Béla levelei* [Béla Bartók Letters], ed. János Demény (Budapest, 1948, 1951, 1955, 1971, 1976). In original Hungarian or Hungarian translation.
BBrCal	'Bartóks Briefe an Calvocoressi (1914–1930)', ed. Adrienne Gombocz and László Somfai, *Studia Musicologica*, 24 (1982), 199–231. In original French, German, and English.
BL	British Library.
BL Add. MS	British Library, Additional Manuscript.
*DB.*i–vi	*Documenta Bartókiana* (Budapest, 1964, 1965, 1968, 1970, 1977, 1981).
DM	*Daily Mail*
DT	*Daily Telegraph*
ES	*Evening Standard*
KBKK	'Korrespondenz zwischen Bartók und der holländischen Konzertdirektion "Kossar" ', ed. János Demény, *Documenta Bartókiana*, 6 (1981), 153–229. In original German.
MB	*Music Bulletin*
MG	*Manchester Guardian*
ML	*Music and Letters*
MMR	*Monthly Musical Record*
MNH	*Musical News and Herald*
MO	*Musical Opinion*
MP	*Morning Post*
MQ	*Musical Quarterly*
MR	*Music Review*
MS	*Musical Standard*
MT	*Musical Times*
MTA BA-B	Hungarian Academy of Sciences, Institute of Musicology, Budapest Bartók Archive collection.

MTA BH	Bartók Estate collection, housed in the Budapest Bartók Archive.
NHQ	*New Hungarian Quarterly*
RT	*Radio Times*
ST	*Sunday Times*
SzFAC.I–IV	Szathmáry Family Archives, Chicago, Bartók/Arányi collections I–IV.
VBBH	'Vier Briefe Bartóks an Philip Heseltine', *Documenta Bartókiana*, 5 (1977), 139–41. In original French and English.
ZT.ii, iii, vii, x	*Zenetudományi tanulmányok* [Studies in Musicology] (Budapest, 1954, 1955, 1959, 1962).

Part I

ON TOUR

I.

A FALSE START

[Dec. 1905]

My biographical information is as follows:

B. 1881. Nagyszentmiklós. At 7, lost my father; my mother (*née* Paula Voit) brought me up under very difficult conditions.

Aged 10, Nagyszöllős; Mr. Altdörfer discovered my talent. First my mother taught me, then László Erkel in Pozsony, then I went on to the Budapest Academy of Music (Thomán, Koessler). My *Kossuth* symphony was performed and warmly acclaimed in Budapest in 1904; and later, in Manchester. In March this year, in Budapest, I scored a success as a pianist; also two weeks ago in Manchester. A week ago my orchestral suite, in all its Hungarianness, caused a sensation *in Vienna*.

I have never written anything like this about myself before, but I can't help that—it is just these 'successes' that are the most important points in a curriculum vitae.

With patriotic greetings,

Yours sincerely,

Béla Bartók[1]

At the age of 24 Béla Bartók could be satisfied with the course of his career. Since graduating from the Budapest Academy of Music in 1903 he had consolidated a reputation for sensitive pianism and patriotic, if not fully matured, composition. Despite occasional setbacks his performances in such centres as Berlin, Budapest, Vienna, and Paris had marked him out as yet another promising pianist-composer in the Romantic mould of a Chopin or Liszt. Had Bartók decided to play out that role, as did his compatriot Ernő Dohnányi, a comfortable future would probably have been his lot. But his mind was of an altogether more radical cut. As he shrugged off the enthusiasms and *naïveté* of late adolescence, his interest in folk-music grew; under its influence, his compositional style became more radical; his dedication to a pianistic career wavered; his popularity declined at home and abroad. By 1913, in the face of charges of musical barbarism and anarchy, he had become a recluse, and only re-emerged onto the international stage in the 1920s.

While most of Bartók's British tours occurred in the 1920s and 1930s, his first two visits took place in the few years of 'false start' to his career which immediately followed his graduation. These early visits were undertaken through the initiative of Hans Richter, the famous exponent of Wagner's music, who had since the turn of the century been the conductor of Manchester's Hallé Orchestra. Richter had been born in the western

---

[1] Biographical notes in the form of a letter, Bartók to Géza Mayer, n.d. [*c.* 5 Dec. 1905], in Hungarian, reproduced in English in *BBL* p. 54.

Hungarian town of Győr (Raab) in 1843, and had during many years of absence abroad maintained an intense affection for his homeland. While in Hungary on a visit during June 1903, Richter was introduced to Bartók by Károly Gianicelli, a professor at the Budapest Academy.[2] Gianicelli particularly wanted Richter to hear the symphonic poem *Kossuth*, on which Bartók was then working, and hopefully to accept it for performance. This work, written in the style of Bartók's current idol, Richard Strauss, depicted events in the abortive Hungarian War of Independence of 1848–9, in which Lajos Kossuth had played a leading part. Its ten sections expressed the hero's anxiety at the dangers besetting the homeland, his call to arms to the Hungarian people, and the final triumph of the Austrian oppressors through their superior numbers. The ending would have stirred the heart of anyone sensitive to Hungary's plight: 'All is finished. Hungary lies in deepest woe, in deepest mourning—A hopeless silence reigns'.[3] Richter was indeed moved by the work, as Bartók afterwards reported to his mother:

Well, I wasn't expecting such a splendid result. And what's most pleasing is that he liked my composition so much, and that of all works it was *Kossuth*. I am pleased not only for my own sake, but also because a completely *Hungarian* piece of music will be performed in England—Hungarian in its subject, Hungarian in its style, in short Hungarian in every respect—a piece which glorifies our greatest patriot and which gives voice to our feelings against Austria. Among others, the 'Gotterhalte' section pleased Richter especially. While I was playing he said: 'Bravo, grossartig'. When I had finished he even remarked that that section was antidynastic, and that he was very pleased. (Because his wife's family had suffered a lot after the War of Independence.) At once Richter drew up the programme for the concert in Manchester . . . He can only perform 'Kossuth' if it is well scored. We hope that it will be. Richter's daughter was there as well; she said goodbye with the words 'See you in Manchester'.[4]

Pending confirmation of Richter's promises, Bartók worked hard on the orchestration of *Kossuth*, as only a piano version had yet been completed.[5] By the end of July 1903 Richter had honoured his promises: Bartók was engaged for 18 February of the following year and even invited to stay with the Richters while in Manchester.[6] His only reservation concerned Richter's determination that he should play as a piano solo the *Variations on a Theme by Handel* by the Budapest composer Robert Volkmann. Bartók did not think the set very effective.[7] After the concert he would have reason to recall these doubts.

The commitment from Richter raised Bartók's spirits. This was a definite professional engagement, in fact his only firm booking, apart from a concert which he had himself arranged for Berlin in December.[8] Even there

---

[2] *BBcl* pp. 102, 105–6.   [3] *BBE* p. 403. See also *BBcl* p. 103.
[4] Letter, Bartók to his mother, 27 June 1903, *BBcl* pp. 106–7. See also *Pressburger Zeitung*, 1 July 1903.   [5] *BBL* p. 25.
[6] *DB*.iii pp. 32–3.   [7] *BBlev*.v pp. 56–7.   [8] Ibid. 63.

Richter's support proved useful. When recently in Manchester, the composer-pianist Ferruccio Busoni had heard good reports of Bartók from Richter, and so curiosity led him on his return to Berlin to attend Bartók's performance. The young Hungarian was touched when, after a bracket of his own compositions, Busoni came backstage and congratulated Bartók on his work, above all a *Fantasie* (1903) which he had just played.[9] Richter's prompt acceptance of the merits of *Kossuth* also helped Bartók's advancement in another way. The Budapest Philharmonic Society was now persuaded to mount the work before the Manchester performance. On 13 January 1904 *Kossuth* was premièred in Budapest. The growing mood of nationalism ensured a generally favourable reception, at least from the Hungarians in the audience. Bartók's mother, who was present, recalled to a family friend some months later: 'I'll never forget the thunder of the applause as Béla went on bowing, looking so happy (he was called 8 or 10 times), and as for me, I wept for joy.'[10]

Buoyed up by these recent successes, Bartók left for England around 10 February 1904. After numerous changes of train he finally arrived at Richter's home in Bowdon, just outside Manchester. As it was his first journey beyond central Europe, Mrs Richter rather mothered him, and even sent a short letter to Bartók's mother to let her know that her son had arrived safely and was most welcome.[11] In Bartók's first full letter home— he had sent five postcards to his mother during the trip—he characteristically showed most concern for questions of money and travel arrangements. There was only brief mention of the Richters, and nothing at all about music:

I have been somewhat surprised to find that life is not as expensive as I had expected. In London 'cabbies' charge 2/2 (= 1.40 Kr), whereas in Vienna they ask 1.60 Kr. Even porters are content with 30 kreutzers (3d) (d = penny). I've got used to the English money and counting very quickly. The battered copper coins (1-penny pieces), as big as this [Bartók drew a circle] are very quaint. There is only 1st and 3rd class from London to Manchester. (As regards comfort, there's nothing to choose between them.) I travelled 3rd for 16 shillings; then there was 2/2 for lunch and 8d for a Hungarian Apollinaris I had with it; I gave 3d as a tip.[12]

Bartók appears to have spent the few days before the concert practising, attending rehearsals, visiting a museum, and hearing his friend Ernő Dohnányi (who had also enjoyed Richter's support) perform his own *Passacaglia*, Op. 6 in a local concert.

At the time of Bartók's visit, Manchester was the leading provincial centre

[9] *BBL* p. 35.
[10] Letter, Bartók's mother to Mrs Gyula Baranyai, 4 Apr. 1904, in Hungarian, reproduced in English in *BBL* pp. 39–41. For the collected reviews of the concert see *DB.i* pp. 30–59.
[11] *ZT.ii* p. 441.
[12] Letter, Bartók to his mother, 12 Feb. 1904, in Hungarian, reproduced in English in *BBL* pp. 37–8.

of music in Britain. Charles Hallé had died only nine years previously
leaving a legacy of perhaps the finest orchestra in the country and a newly
founded music college. In the following decade, musical standards were
further increased by Richter's work as conductor and the tireless propa-
gandizing of the *Manchester Guardian*'s music critic, Arthur Johnstone. In
inviting Bartók to take part in a Hallé concert, however, Richter was
behaving atypically. He had a reputation for cold-shouldering contemporary
works, so much so that there was some dissatisfaction among patrons at the
narrowness of the concert diet.[13] One 'modern' whose works Richter was
readily prepared to conduct was Richard Strauss, and the obvious affinity
between *Kossuth* and Strauss's *Ein Heldenleben*, coupled with the issue of
national sentiment, seems to have swayed Richter towards Bartók's case.
The Manchester critics were, therefore, especially interested to witness
Bartók's début and to hear his much-advertised composition. An advance
notice of the concert had even stated, quite incorrectly, that Bartók had 'not
yet reached man's estate', thereby adding a sensational touch to the event.[14]
Bartók certainly featured highly in the programme on 18 February:[15]

PART I
Symphony 'Unfinished'   Schubert
Spanish Rhapsody (scored by F. Busoni)   Liszt
    Mr. Béla Bartók
PART II
Symphonic Poem 'Kossuth'   Béla Bartók
Variations on a Theme by Handel   Volkmann
    Mr. Béla Bartók
Suite for Orchestra Op. 39   Dvořák

As was to be expected, the novelty of this programme caused comment
among the critics: 'Last night's programme was, for two reasons, a most
unusual one', began the *Daily Dispatch*'s review; 'to begin with, we had no
fewer than three pieces all for the "first time at these concerts," and,
strangest innovation of all, the Symphony was not played last, but occupied
the premier place on the programme.'[16]
    Bartók's first exposure to the English critics was not a comfortable

[13] See C. B. Rees, *One Hundred Years of the Hallé* (Norwich, 1957), 45; Michael Kennedy,
The Hallé Tradition (Manchester, 1960), 136; *BBcl* pp. 120–1.
[14] Anon., 'The Hallé Concert', *Manchester Courier*, 12 Feb. 1904.
[15] MTA BA-B 2049.
[16] Reviews of this concert mentioned: S. B., 'The Hallé Concerts', *Daily Dispatch*, 19 Feb.
1904; Anon., 'Music', *Manchester Weekly Times*, 26 Feb. 1904; Anon., 'The Hallé Concerts',
Manchester Evening News, 19 Feb. 1904; Anon., 'The Hallé Concert', *Manchester Courier*, 19
Feb. 1904; F. W. G. B., 'Music', *Manchester City News*, 20 Feb. 1904; Anon. [Arthur
Johnstone], 'The Hallé Concerts', *MG* 19 Feb. 1904; Anon., 'Music in Manchester', *MT* 45
(1904), 188; Anon., 'Our Contemporaries', *MS* 19 Mar. 1904, p. 190.

experience, but as a pianist, at least, he was enthusiastically received. 'Exceptionally rich and expressive tone', recorded the *Manchester Weekly Times*; 'as a pianist we shall no doubt hear further of Mr. Bartók', wrote the critic from the *Manchester Evening News*. The review in the *Manchester Courier* elaborated Bartók's pianistic virtues: 'Mr. Bartók's style at the piano is very good; he plays with the earnestness of an enthusiast, yet at no time falling into exaggeration. His technique, moreover, is superior to all difficulties.' Needless to say, not all the reviews were as ecstatic. The *Manchester City News*'s correspondent found Bartók's keyboard mannerisms irritating. Arthur Johnstone of the *Manchester Guardian*, while happy with Bartók's performances, questioned the inclusion in the programme of Volkmann's *Variations*, which he found 'very peculiar'. The *Musical Times* was more blunt about the programme. Despite several recalls and an encore after the Volkmann piece, its critic would write of the evening: 'The bulk of the Hallé audience cared for nothing in the sixteenth concert except the "Unfinished" Symphony, which was poetically and faultlessly rendered. Those who waited for the Dvořák Suite at the end no doubt accepted it as fairly satisfactory music; but the intervening pieces were all of a kind which the Manchester public does not like.'

With *Kossuth* the critics were less generous. They agreed that the symphonic poem was not a great composition, and differed only in the degree of offence which they felt the work had given. The section depicting the battle between the Hungarians and the Austrians was recalled in nearly every review. Bartók's programme-notes, included in the concert booklet, vividly described the music's plot at this point:

Then is heard the sound of the enemy's host, approaching nearer and nearer. It is characterized by the motive of the Austrian hymn (*Gotterhalte*)—The armies join in battle—Assault after assault is made. At last the superior numbers of the enemy triumph. The catastrophe comes (*fff* on tympani and tam-tam). Only a few of the Hungarians, who have survived the conflict, fly before the vengeance of the victors.[17]

Under the sub-title 'Strauss Out-Straussed', the *Daily Dispatch* gave a detailed account of the work, appending the admonition:

One cannot allow such a cacophonous display as is presented in the 'Battle' section to pass without a word of remonstrance and regret. Take all the demons of Berlioz, Strauss, and Elgar put together, and multiplied 'ad infinitum'; over that crude mixture, imagine a minor version of the Austrian hymn played now on the Contra bassoon, and then on trumpets and trombones! All of course fortissimo. The result would be painful, if it were not so laughable. . . . it is a pity that the young Hungarian has allowed his patriotic feelings, coupled with a desire to make his work too sensational, [to] run away with him.

[17] *BBE* pp. 402–3. See also *ZT*.ii p. 422.

A very lengthy review of the work in the *Manchester Guardian* attracted so much interest that it was reprinted the following month in the 'Our Contemporaries' column of the London journal *Musical Standard*. In this review Arthur Johnstone baldly stated: 'If anyone thinks that it [*Kossuth*] really pleased the Manchester public he is under a very gross delusion.' Johnstone objected in particular to the use of the Austrian hymn in the 'Battle Scene': 'apart from the function of the theme as a barefaced label, analogous to a piece of writing in the middle of a painted picture, "Yankee Doodle" would have answered the purpose much better'.

The conservative critic of the *Manchester Courier* found the very direction of Bartók's composition the cause of his troubles. Citing the 'demons' of Wagner, Liszt, and Strauss (a different trinity from that identified in the *Dispatch*'s review) as the seducers of Bartók from the true ways of Mozart and Beethoven, he concluded that the great formal freedoms inherent in the symphonic poem had been poorly handled by Bartók, resulting in an aimless, 'not to say repellent', composition. Less colourful, yet more prophetic, were the grounds for criticism put forward in the *Musical Times*. While admiring Bartók's grasp of orchestral resources and Straussian techniques, the correspondent regretted that the work had not contained elements of Hungarian folk-music. Only because of the presence of the composer, and his excellent piano-playing, concluded this critic, had the work been received cordially.

As often happens when an artist is abroad, the reports appearing back home were somewhat less critical. In the *Pressburger Zeitung* of Pozsony (Bratislava), the notice appeared: '[Bartók] presented himself before an English audience which knew absolutely nothing of him and was therefore objective in its opinions. Dr. Hans Richter himself has sent a telegram . . . with the following complimentary words: *Kossuth* was received sympathetically, the young pianist gained a triumph.'[18] Two weeks later in the 'Foreign News' column of Budapest's music publication *Zenevilág* the truth had slipped a little further into the background. Now Bartók had triumphed 'with masterly playing', and 'the English papers wrote about the symphony in glowing terms'.[19]

Bartók, none the less, took the criticisms of his composition most seriously. Despite his sparse English, he discussed the offending Austrian hymn episode with John Foulds, a member of the Hallé, who also suggested that a folk-tune would have been a more appropriate foundation for the section.[20] In a postcard to his Academy piano-teacher, István Thomán, Bartók gave a fair account of his reception: 'I gained a nice success yesterday, mainly as a pianist. "Kossuth" was not liked that much, although

[18] Anon., 'Bartóks Künstler-Erfolg in Manchester', *Pressburger Zeitung*, 22 Feb. 1904.
[19] Anon., 'Külföldi Hírek', *Zenevilág*, 8 Mar. 1904.
[20] John Foulds, *Music Today* (London, 1934), 254.

it was received favourably enough. The critics find fault with it in many regards. I shall send the programme and a few newspaper articles next time.'[21] But somehow the reviews never filtered back to Hungary: '. . . unfortunately I mislaid them somewhere *en route*. It's a pity because there were one or two interesting comments in them', he later wrote to a friend.[22] After the great success of *Kossuth* at its première in Budapest, Bartók was probably embarrassed to have the details of this severe, even mocking, English criticism known back home.

Richter's reaction to the performance and its reviews is harder to ascertain. As an avowed conservative it was unusual for him to be taken to task for mounting too revolutionary a work. The Austrian hymn episode, which he had praised, was now generally condemned; the Volkmann piece, which he had chosen, was considered a novelty of inferior standard. If Bartók is to be believed, Richter chose to retreat from his earlier judgements and to exercise a greater caution towards his young Hungarian friend. Writing a frank letter to his mother soon after leaving Britain, Bartók elaborated on the matter:

Hm! Matters stand strangely with Richter. Before the rehearsals began he was very fond of Kossuth; he even said that he would attempt to have it performed in London next year, and would have me play the piano both there and in Manch. etc.; and even when I played K. to him once again on the piano, he again declared 'ausgezeichnet, famos'. But then, chiefly after the concert, he became frightened or something, and he stopped praising the work as enthusiastically. And he has not said anything definite about what will happen next year. . . . The trouble is, firstly, that Richter's opinions are rather easily influenced; secondly, that one can't take his promises absolutely seriously. . . . Then, it is a difficulty that Richter is not actually idolized in England to the extent that Batka had stated: one of the newspapers has spread the rumour that there will be no Richter concerts in London next year, because they finished up with a deficit this year. The fact is that Londoners are not content with Richter's programmes. He does not introduce enough new items; everything runs according to the old routine.[23]

(By this time the Richter London Concerts had been a feature of the metropolis's musical life for twenty-five years. The conductor himself realized that changes were needed and eagerly fell in with plans for the formation of the London Symphony Orchestra in the early months of 1904. He conducted its first concert on 9 June 1904 and maintained a close relationship with the orchestra, alongside his Hallé commitment, until his retirement from conducting in 1911.[24])

[21] n.d. [19 Feb. 1904], *BBlev*.v p. 75. See Plate 2. See also *BBlev*.v p. 76.
[22] *BBL* pp. 38–9.
[23] Letter, Bartók to his mother, 7 Mar. 1904, *BBcl* pp. 120–1.
[24] See Hubert Foss and Noel Goodwin, *London Symphony* (London, 1954), 1–30, and Reginald Nettel, *The Orchestra in England: A Social History* (London, 1946), 217–29.

Although not through Richter's efforts, Bartók did gain another opportunity to perform in Manchester during his visit of February 1904. This chance was welcome, because having come as an 'unknown artist' he had not been paid a fee for his Hallé appearance.[25] It came about through the advice of Dohnányi, who had gained some familiarity with British musical circles and could thereby help to advance his friend's cause. On 10 February he informed Bartók that there might be a place for him on 20 February in the programme of one of the Ladies' Concerts, which the piano firm of John Broadwood and Sons had recently instituted as a counterpart to the long-established Gentlemen's Concerts.[26] These Saturday afternoon gatherings were intended to suit the tastes of ladies only a few years out of the Victorian era, and therefore, consisted of many light items, such as songs and shorter works for piano or violin.

Never in the slightest a 'popular musician', Bartók none the less wrote to Broadwood putting forward a selection of works. In reply, the organizers declined his offer of Schumann and Liszt sonatas, saying that they were 'rather too long', and chose instead the less substantial pieces on his list.[27] Even so, his contribution did not blend well with the rest of the programme. In the first half his Chopin Ballade was sandwiched between vocalists singing 'Love's Philosophy' by W. Henri Zay and 'Love the Pedlar' by E. German. After the interval, Bartók returned to play his own *Fantasie*, 'Aufschwung' from Schumann's *Fantasiestücke*, and a Liszt–Paganini study.

The reviews of these performances before the ladies were generally brief. Only the *Manchester City News* afforded undiluted praise, finding the playing of all works very effective.[28] For the *Manchester Guardian*'s writer, too, Bartók merited praise for an artistic and effective presentation, although this critic regretted the considerable memory lapses which occurred in both the Chopin and Liszt pieces. Bartók's *Fantasie* caused some disagreement among the critics. It was 'a piece in a meditative vein, original in harmony and colouring and decidedly attractive', according to the *Guardian*. The *Daily Dispatch*'s correspondent, however, received 'an impression of vagueness—as if the composer had something to say but did not quite know how to express himself'. Words of caution were offered by the generally sympathetic critic from the *Manchester Courier*: the compositional direction pursued in the *Fantasie* might 'lead to only barren results'. Compared with the first Manchester performance Bartók considered this second appearance of little consequence. It is only touched upon in his

[25] DB.iii pp. 32–3.
[26] MTA BH 388. See also DB.iii pp. 38–40.
[27] MTA BH 217.
[28] Reviews mentioned: Anon., 'The Ladies Concerts', *Manchester City News*, 27 Feb. 1904; A. J. F., 'The Ladies Concerts', *MG* 22 Feb. 1904; Anon., 'The Ladies' Concerts', *Daily Dispatch*, 22 Feb. 1904; Anon., 'Manchester Music', *Manchester Courier*, 22 Feb. 1904. A further review appeared in the *Manchester Weekly Times*, 26 Feb. 1904.

correspondence,[29] and does not feature in the list of concerts proudly compiled by his mother.[30]

The concerts over, Bartók stayed on with the Richters for a further six days before going to London. When leaving on 26 February he presented them with an elaborate card produced by a Budapest photographer, on which he wrote a simple message of remembrance.[31] For many years the card remained in the possession of Richter's youngest daughter, Thildi, to whom it would seem the visit meant most. During the fortnight of Bartók's stay they had developed a friendship which was to be maintained by an intermittent correspondence and several meetings over the following few years.[32]

In London Bartók spent a busy time, staying, probably along with Dohnányi, at the home of Mrs Oliverson, a leading patron of the arts. To his mother he wrote of these London days: '[Mrs Oliverson] has invited me to stay again, so long as she is still then living in London (Oh, yes!) . . . I really lived like a lord in London—I have never been driven about as much as I was during those six days (and mostly at somebody else's expense).'[33] While in the capital Bartók showed his recent compositions to a number of musicians of Dohnányi's acquaintance. Several, including the pianist Fanny Davies (one of Clara Schumann's last pupils), expressed a desire to perform his works.[34] But most importantly for the future, the Broadwood concert organizers had, at the prompting of Dohnányi, promised Bartók six engagements for the following year, including a London recital.[35] With such an assurance Bartók must have departed from Britain with some feeling of satisfaction. Despite a good deal of adverse criticism, his first concert tour of Britain had succeeded in arousing interest in his activities, both as composer and pianist. Of additional comfort was the knowledge that on 29 February the Budapest Opera Orchestra had successfully premièred his Scherzo (1902).

Later in 1904, through the mediation of Thildi Richter, Bartók met her father at Bayreuth and tried to interest him in his new Scherzo, Op. 2 for piano and orchestra.[36] Richter's response was quite different from that of the previous year when he had encountered Kossuth:

[29] BBlev.v p. 75.

[30] Mrs Béla Bartók, sen., 'Béla hangversenyeinek jegyzéke', unpublished manuscript in the possession of Béla Bartók, jun. (Budapest), p. 1. Bartók's mother does, however, list two other concerts in Manchester, one on 24 Feb. (which may have been a private affair) and the other in March (which is certainly incorrect).

[31] From a copy in the private collection of János Demény (Budapest). Present location of original unknown.

[32] See six unpublished letters, Thildi Richter to Bartók, dating from 20 Mar. 1904 to 23 Mar. 1906, in German, MTA BH 2138–43. It is clear that the correspondence continued further (BBcl p. 162). Bartók renewed contact with Thildi Richter (by then, Thildi Loeb) during his British tour of 1922.

[33] Letter, Bartók to his mother, 7 Mar. 1904, BBcl pp. 120–1.

[34] Ibid.   [35] BBL pp. 38–9.   [36] MTA BH 2138–9.

I showed it [the Scherzo] to Richter in Bayreuth; he said it was '*ein gelungener musikalischer Scherz*'. The only thing he objected to was the title. He said the piece was too grandiose, too complex, too 'sparkling' for such a plain title. Then he added: 'You mustn't expect it to be generally liked, though'. And so, even though Richter liked it, the result is nil.[37]

Despite Richter's caution over this particular piece, he was prepared to give Bartók a second chance with the Hallé. Writing to Bartók for his twenty-fourth birthday, in March 1905, Thildi Richter confided:

Professor Gianicelli wrote to me telling how wonderfully the 'Todtentanz' by L[iszt] was played by you, which pleased me very much. I was less pleased that your Scherzo could not be played owing to its difficulties! Why don't you ever write something which can be played without danger to life and limb? Nobody would be more happy than me, because I honestly admire your style and talent, and wish you a resounding success. . . . Papa expressed a very nice opinion of you. I am sure that you will be pleased to hear that. He said that you showed the extremely promising talent of a genius; then again, he said yesterday that you will probably get an engagement in the coming season for a Hallé concert. That would be fine, wouldn't it?[38]

By June Bartók had formally been invited for an engagement on 23 November 1905.[39] His role was restricted to that of pianist playing items by established composers. More depressing, however, was the abrogation by the Broadwood firm of its promise of engagements. Bartók, none the less, continued to dream of these concerts: 'The Broadwood people, anyway, are still bound by their promise to me, which they could not honour, or for certain reasons did not wish to honour, this year. Hence, my confidence in England. In any case it is good that I have something to cling to in the event that absolutely nothing should turn out right.'[40]

In Vienna, late in October, Bartók began to make travel arrangements and to take lessons to revise his English. He planned to leave for England on 8 November, and to return to Vienna by the end of the month for the first performance of part of his *Suite* No. 1, Op. 3. Since his recent lack of success in the Rubinstein Competition in Paris, however, he had started to take his career much more seriously, and was quite prepared to stay longer in England if extra engagements could be secured.[41] But despite his efforts, he was to be back in Vienna by the end of the month: last-minute approaches were made to the firms of E. L. Robinson, Bechstein, and Broadwood by Bartók and several of his London friends, but the responses were uniformly negative.[42]

By way of contrast with these professional frustrations, the Hallé concert

---

[37] Letter, Bartók to István Thomán, 18 Sept. 1904, in Hungarian, reproduced in English in *BBL* p. 43. The first performance of Bartók's Scherzo, Op. 2 finally took place in 1961.

[38] Unpublished letter, Thildi Richter to Bartók, 25 Mar. 1905, in German, MTA BH 2140.

[39] *BBlev*.v p. 88.

[40] Letter, Bartók to István Thomán, 21 Apr. 1905, *BBlev*.v p. 87.

[41] *BBcl* p. 143.   [42] MTA BH 217, 925, 2158.

on the evening of 23 November was a decided success. In a more adventurous programme than most in the series, Bartók played Liszt's *Totentanz* for piano and orchestra, and as a solo item, Bach's *Chromatic Fantasia and Fugue.* The remainder of the concert consisted of some of Richter's favourite works by Brahms, Richard Strauss, and Beethoven. From the many critics present, Bartók's playing of Bach elicited complimentary, even rapturous reports. 'Just the right Bach spirit', commented the *Musical News.*[43] The *Musical Standard*'s Manchester correspondent found the exposition of the fugue 'capital' and noted that such an important work had not been heard at a Hallé concert since 1870. Ernest Newman, who had recently succeeded Arthur Johnstone as chief critic for the *Manchester Guardian,* found some small difficulty with the rendition, as he explained in what would be the first of many reviews of Bartók's concerts: 'The performance of the Fugue did not, perhaps, quite grow enough in power towards the end . . . but it was always lucid, while the Fantasia was beautifully played throughout.' The *Manchester Weekly Times* wrote even more appreciatively of the performance: 'It is not often that a pianist in the Free Trade Hall can cause each member of the audience to feel that the message is for him alone, yet this is what Mr. Bartók did. The sensation of being in a vast space was lost, so effectually did the music penetrate to every corner.'

The prevailing British hostility to Liszt's music ensured a rougher passage for Bartók's performance of the *Totentanz.* In an unfettered exercise of free expression the newspaper reviewers waged war over the value of the work, the appropriateness of its performance at this concert, and the merits of Bartók's rendition. 'Why did Liszt write such a paraphrase on the "Dies Irae"?', began the *Daily Dispatch*'s harangue. 'Why did Mr. Béla Bartók play it at last night's concert?' Bartók did his best with this ugly work, the review continued, but should have chosen an entirely different work. This sentiment was echoed in the *Manchester Courier.* It was left to Ernest Newman to look beyond the unappealing surface of the piece and to consider Bartók's keyboard technique in some detail:

Liszt's 'Todtentanz', regarded as mere music, is often rather tiresome. . . . A good half of his time Liszt is palpably playing at being diabolical, telling us, as the Fat Boy told Mr. Pickwick, that he wants to make our flesh creep, but scarcely ever raising a decent horripilation. As a show piece for a pianist, however, it is magnificent. Last night Mr. Béla Bartók played it quite superbly. Now and then he was swamped by the orchestration, but that was Liszt's fault, not his. He gets a big and virile tone out of the piano, reminding us somewhat of the quality that Frederic Lamond draws

---

[43] Reviews mentioned: F. B., 'Provincial', *Musical News*, 2 Dec. 1905, p. 493; W. H. C., 'Music in the Provinces', *MS* 9 Dec. 1905, p. 376; E. N., 'The Hallé Concerts', *MG* 24 Nov. 1905; Anon., 'Music', *Manchester Weekly Times*, 1 Dec. 1905; S. B., 'The Hallé Concerts', *Daily Dispatch*, 24 Nov. 1905; Anon., 'The Hallé Concerts', *Manchester Courier*, 24 Nov. 1905.

from it. He was remarkably good in passages where the tone has at the same time to be sustained and gradually diminished in power; his pedalling was evidently very expert.

At the end of the performance of this work there was an eerie silence before the audience's applause broke. 'This may have been due to a unanimous sigh of relief,' suggested the *Daily Dispatch*; 'when the applause did come, it was doubtless evoked more by the performance than the piece.' The *Musical Standard*'s correspondent implied a more favourable reason: amazement at the marvellously 'liquid quality' of Bartók's playing.

Bartók's response to his second visit to Britain is difficult to ascertain. Although he sent three picture postcards to his sister, these contained no personal or musical impressions—merely birthday greetings and comments about London, couched in the uninformative tone of elder brother to younger sister.[44] Only after returning to Vienna, in a letter to his friend Emma Gruber (later Kodály's wife), did Bartók give some news of his British activities.[45] He confessed that the trip had not really been a success, despite the favourable reception in Manchester. The family of the Hungarian violin virtuoso Ferenc Vecsey had tried hard to advertise his talents in London, but to no avail.[46] Bartók had even performed a small favour for J. A. Fuller-Maitland, the chief music critic of *The Times*, but again to no immediate advantage.

Richter's reaction to Bartók's second visit is even harder to judge. He wrote a pleasant postcard to Bartók during the following month,[47] but his opinion that the Hungarian had taken a turn in the wrong compositional direction was undoubtedly strengthening. His true feelings emerged two years later in a letter written to Károly Gianicelli, the professor who had originally introduced Bartók to him. In response to a suggestion that he perform Bartók's *Suite* No. 1, Op. 3 Richter replied: 'Well, come now, dear professor! What do you think? Not even a run through!!! If Bartók's suite is a "crazy" work then I'm not inclined to perform it. And assuredly it is crazy, when I consider that Bartók was capable of memorizing [Strauss's] Heldenleben.'[48] Another factor probably contributed to Richter's unfriendly attitude: according to Richter family sources, Bartók wished to marry Thildi, and her dismissal of his serious advances contributed to a general cooling in relations around this time.[49]

Bartók's first patron in Britain no longer wished to perform his works or

---

[44] Postcards of St Pancras Station, High Street Haslemere, and Hyde Park Hotel, *BBcl* pp. 143–4.   [45] *BBlev*.v p. 723.

[46] In 1905 Vecsey was only 12, but had already performed frequently with the London Symphony Orchestra. Sibelius had indeed dedicated his Violin Concerto (1903, rev. 1905) to him. In 1906 Bartók toured Spain and Portugal with Vecsey.   [47] *DB*.iii p. 41.

[48] Letter, Richter to Gianicelli, 24 Sept. 1907, in German, held in the Archive of the Hungarian State Opera, partially reproduced in *ZT*.ii pp. 376–7 and *BBlev*.iv p. 12.

[49] Unpublished letter, Eleonore Schacht to the author, 11 Oct. 1986.

to schedule his performances. Nor could Bartók gain alternative engage-ments with other concert organizations. His British trail had run cold and would remain so for over sixteen years. By then so much would be different: his compositional style, his repertory, and his reputation. But the country which he visited would also be different, freed from the many stultifying effects of Victoriana and shocked by the war into a more open-minded attitude to the arts.

2.

A GLIMMER OF RECOGNITION

The announcement of Bartók's visit to Britain in 1922 generated great interest in musical circles. This interest owed nothing to his early visits of 1904–5, but was the result of a more recently inspired awareness among British musicians of Bartók's position in the vanguard of contemporary music in Europe. The seeds of that awareness had been sown as early as 1911 by occasional articles, reviews, and performances of Bartók's music. In an age of growing British attention to folk-music—fostered by such musicians as Sharp, Grainger, and Vaughan Williams—Bartók's folk-inspired music aroused at least a modest curiosity. These pre-war promotions of his music were undertaken quite independently from Bartók himself, who was by this time withdrawing from musical life both at home and abroad. Even the First World War, with its strictures on the performance of music by enemy aliens, was not able entirely to prevent the spread of interest in his music in Britain. From 1920, largely through the activities of the critics Calvocoressi, Heseltine (Warlock), and Gray, a flurry of articles and performances ensued, paving the way for Bartók's visit of 1922.

Because of the lack of direct contact with the composer, early performances of Bartók's music in Britain are difficult to trace. Writing in 1922, the critic Edward J. Dent recalled a performance of the 'Bear Dance' (1908) by O'Neill Phillips, a young pupil of Busoni, soon after its composition: 'Fifteen years ago the "Bears' Dance" [sic] seemed hardly music at all. Its composer had struck definitely away from the romantic tradition.'[1] Around this time, too, Bartók's easier piano pieces—the *For Children* series (1908–9) and *Ten Easy Pieces* (1908)—were starting to be used by a few progressive music teachers, including Colin Taylor at Eton.[2] Academic consideration of Bartók's works followed somewhat later. During 1911–12 A. Eaglefield Hull, a musician with special interest in contemporary music, presented a series of lectures in Lancashire and Yorkshire dealing with Bartók's music.[3] At this time, too, occasional short reports mentioning Bartók's activities in Continental Europe began to appear in the music press,[4] leading in January 1912 to the first review article, in the *Musical Standard*.[5] Writing in the journal's 'Foreign Musical

[1] 'Music', *The Nation and the Athenaeum*, 1 Apr. 1922, p. 32.
[2] See Colin Taylor, 'Peter Warlock at Eton', *Composer*, no. 14 (Autumn 1964), 9–10.
[3] See MTA BA-B 3477/90.
[4] See, e.g., 'Foreign Notes', MT 52 (1911), 476, 741–2; 53 (1912), 536; 54 (1913), 339.
[5] MS 27 Jan. 1912, pp. 55–6.

Reviews' column, Mrs Franz Liebich described the activities of Bartók and Kodály as collectors of folk-music, and outlined the mixture of folk and modern, particularly French, elements in their compositions. Her very general conclusion expressed an optimism certainly not felt by the composers themselves at this time: 'their music is extraordinarily interesting and, as they possess a keen sense of beauty, it is also melodically beautiful. Music being as natural to the Hungarian as the air he breathes, it follows that the renaissance of the art in Hungary is having remarkable results.'

The following year another critic undertook to advertise the work of the 'New Hungarian School' more systematically. This was M.-D. Calvocoressi, a resident of Paris, who was then best known for his advocacy of Russian music. Calvocoressi remembered Bartók's contribution to the 1905 Rubinstein Competition in Paris and considered that the decision not to award him a prize had been unjust.[6] His article 'Perplexities of the Modern Music Lover' in the issue of the *Musical Times* for March 1913 concluded with mention of the difficulties experienced in listening to the works of Bartók and Kodály.[7] Two months later, in the same journal, Bartók assumed greater prominence in 'The Problem of Discord', and a short excerpt from the second of the seven *Sketches*, Op. 9b was included.[8] In November 1913 Calvocoressi again reviewed Bartók's music, providing two brief extracts from *For Children* in his essay 'Folk-Song in Modern Music'.[9] Strangely, in this very month, Bartók ventured an enquiry about critical activity in England. He wrote to a friend in London in the hope of learning more about the work of the leading critic Edwin Evans, in particular concerning his writings on modern Russian opera. Bartók did not pursue the matter, however.[10]

1914 was the year in which Bartók's music moved more decidedly into the current of British music-making and criticism. On 11 March Franz Liebich organized a concert featuring Debussy's Clarinet Rhapsody (1909–10), Kodály's Cello Sonata, Op. 4 (1909–10), and his own rendition of several of Bartók's piano pieces dating from 1908–10.[11] 'Ripieno', a critic for *Musical Opinion*, found Bartók's pieces unexceptional: 'Some piano pieces by Béla Bartók also proved to be interesting, although not violently demanding repetition. One thanks Mr. Liebich, however, for his fearlessness and his serious attitude towards works and composers of tentative merit. It

---

[6] See M.-D. Calvocoressi, 'Béla Bartók', *MNH* 11 Mar. 1922, p. 306.
[7] *MT* 54 (1913), 158–61.   [8] Ibid. 300–2.   [9] Ibid. 716–19.
[10] See MTA BH 420.
[11] This was probably not the first concert at which Liebich presented works by Bartók and Kodály. A short review of Liebich's career (*MS* 19 July 1913, pp. 56–7) mentions modern Hungarian music as one of his particular interests. Referring to a photograph of Liebich and his dog, the final paragraph of the article reads: 'The fox-terrier in the photograph has a keen and discriminating ear. He abhors modern composers . . . the Hungarians' music elicit[s] prolonged and piteous howls.'

is to be hoped that he will continue the good work . . .'[12] Another critic, from the *Musical Times*, was more impressed, and wrote that the Hungarians' pieces 'showed that a national idiom and free modern treatment can unite with excellent results'.

In the newspapers and journals of this year Bartók's music came in for increasing comment, particularly in the writings of Leigh Henry, a Briton then residing in Florence.[13] In the *Daily Telegraph* he called for the freeing of musical studies from outworn formulae, citing the works of Bartók and Kodály, amongst others, as models of a truly contemporary treatment of themes.[14] During May 1914 Henry was responsible for the first full-length British article on Bartók, which appeared in the *Egoist* as part of a series entitled 'Liberations: Studies of Individuality in Contemporary Music'.[15] Henry saw Bartók's purpose as the harnessing of Hungarian thought with that of the wider world, so as to provide a broader scope for the development of the qualities inherent in his race. After a cursory examination of several of Bartók's works written between 1907 and 1910, Henry summarized his argument:

The Magyar idiom, always apparent in the works of Bartók, is never treated as an end in itself, but always as a subject for investigation and analysis. As Schönberg by the projection of concentrated introspection enables us to appreciate individual psychology, so Bartók, by persistent examination of the racial characteristics present in his own personality, enables us to estimate the forces controlling national feeling which, impossible to connect and correlate when viewed as a turbulent, emotional whole, admit of full analysis and interpretation by comparison with individual thought and experience.

Some readers, understandably, found Henry's point hard to follow, especially since they knew virtually nothing of the music of either Schoenberg or Bartók. Having also expressed a similar opinion about Bartók in a recent article on Skriabin, Henry was asked by one reader to elucidate his 'unusual criticism'.[16] He obliged: 'Béla Bartók, working from Magyar modes has evolved an entirely novel and personal method of harmonic treatment therefrom, which has in turn vitalised the modes themselves. These modes are the direct outcome of certain racial character-istics.'[17]

To capitalize on this growing interest in Bartók's work, Sir Henry Wood, London's leading conductor, planned for British premières of three of his

---

[12] Reviews mentioned: 'Views and Valuations', MO 37 (1913–14), 553; Anon., 'London Concerts', MT 55 (1914), 259.

[13] In 1922 Henry sent copies of the articles which he had written in 1914 to Bartók (MTA BA-B 3477/126).

[14] Quoted in 'Views and Valuations', MO 37 (1913–14), 552.

[15] Egoist, 1 May 1914, pp. 167–9.

[16] MS, 11 Apr. 1914, pp. 339–40, and 23 May 1914, p. 493.

[17] Letter to the Editor, MS 6 June 1914, pp. 545–6.

orchestral compositions at the Promenade concerts of 1914: the *Rhapsody*, Op. 1 (1904–5) on 20 August, the *Suite* No. 1, Op. 3 (1905) on 1 September, and *Two Pictures*, Op. 10 (1910) on 16 September.[18] Cruellest of fates, war was declared on 4 August. The issue of the correct patriotic response to such foreign works was canvassed among London's musicians. As a first response the Promenade Wagner nights were cancelled, compositions by living German or Austro-Hungarian composers were replaced by more suitably affiliated compositions, and the various Allied National Anthems were played in turn at the conclusion of each concert.[19] Bartók's *Rhapsody* was replaced by an 'Allied' composition, the 'Africa' Fantasy, Op. 89 by Saint-Saëns.[20] But this first policy was considered an overreaction to the situation, and the original 'enemy' works, including Bartók's remaining two, were reinstated. On 1 September, accordingly, the *Suite* No. 1 was performed at the Queen's Hall. Not all members of the orchestra were pleased with the latest policy decision; some objected to performing this strange work, especially in time of war. In *My Life of Music*, Wood recorded the scene (with some benefit of hindsight):

[A. E. Brain] stood up and 'went for' me. 'Surely you can find better novelties than this kind of stuff?' he said indignantly. I saw that there was ⸱ call for a little tact. 'You must remember', I said, 'that I must interpret all schools of music—much that I do not really care for—but I never want my feelings to reflect upon my orchestra. You never know, but I am of the opinion this man will take a prominent position one day. It may take him years to establish it, but his originality and idiom mark his music as the type of novelty our public ought to hear.'[21]

With the war less than a month old most newspapers were too concerned with the German advance on Paris to report mere musical events, but several of the specialist journals did manage to review the performance. According to the *Musical Times*:

Béla Bartók, the young Hungarian progressivist, whose name reached our ears in advance of his music, underwent his first serious trial before the British public on September 1. His Suite in five movements, a comparatively early work, failed to make a deep impression. It is rich in arresting ideas and effects, but the struggle to be interesting is more obvious than the actual interest, and the design and handling lack spontaneity and sense of style. Hungarian national elements impart some value in fact rather than in feeling.[22]

The *Monthly Musical Record*'s review argued along similar lines that Bartók's elaborations upon his subject-matter were out of keeping with that

[18] See *MT* 55 (1914), 513.
[19] See 'Promenade Concerts', *Standard*, 20 Aug. 1914.
[20] See 'Musical Gossip', *Athenaeum*, 22 Aug. 1914.   [21] (London, 1938), p. 380.
[22] Reviews mentioned: Anon., 'The Promenade Concerts', *MT* 55 (1914), 625; Anon., 'The Promenade Concerts', *MMR* 44 (1914), 278–9; Anon., 'Music in London', *MS* 12 Sept. 1914, p. 196. For a less-informed review see 'Concert Notices', *MO* 38 (1914–15), 26.

material's natural spontaneity, 'the result being more creditable to his ingenuity than to his sense of the fitness of things'. A more substantial critique appeared in the *Musical Standard*, where the opinions of Mrs Franz Liebich were paraphrased. After an introduction explaining the differences between gypsy and native Hungarian folk-music, the *Suite* was evaluated:

It is extraordinarily rich in tone colour and glows with vitality, in fact, it has an almost barbaric splendour and power. On the other hand, it is by no means crude, the harmonies are idiomatic and individual, but how far the individuality is racial and not peculiar to Bartók is difficult to say. Mrs. Liebich remarks that the whole work is more or less diatonic, only occasionally foreshadowing the 'pointillisme' of Bartók's later style, an example of which it is to be hoped we may soon have an opportunity of hearing.

That opportunity would not present itself for many years. Performance policy changed again, this time to the detriment of Bartók's third scheduled work, and the embargo remained effective—at least as far as Bartók's works were concerned—for the duration of the war. His exquisite *Two Pictures* were replaced by Joseph Holbrooke's *Imperial March*, a jingoistic work which culminated in a climax based upon snatches of 'Rule Britannia' and 'God Save the King'.[23]

Although the controllers of public music could prevent performances of the works of enemy aliens, they could not easily control other forms of musical expression. Philip Heseltine, who had gained a liking for Bartók's music at Eton and Oxford, continued privately to play and to discuss his music.[24] Frederick Corder, a senior professor of the 'old school' at the Royal Academy of Music, penned a vitriolic article, 'On the Cult of Wrong Notes', which was published during 1915 in the first volume of Schirmer's transatlantic journal, the *Musical Quarterly*.[25] Amazed at the serious attention being paid to such composers as Schoenberg, Skriabin, and Bartók just before the war, he was moved to state: 'now that the entire public has plunged into a hysterical fit of hatred for everything German, one can speak the truth without fear of giving these things their desired advertisement.' The tone of Corder's tirade against Bartók was itself nothing short of hysterical. Bartók was a 'freak composer' and follower of Schoenberg, Corder maintained. He expressed relief that because of the war a stop had been put to 'the production of sheer nonsense', which had become both easy and profitable in the preceding years. 'It almost reconciles one to the awful catastrophe of this European war to think that it will at least sweep away

[23] See *MS* 26 Sept. 1914, p. 227. Wood's score of *Two Pictures*, with the remark 'Proms, Sept 1914' on the title-page, is preserved in the Library of the Royal Academy of Music. See Anne Dzamba Sessa, *Richard Wagner and the English* (London, 1979), 141–2, for a summary of performing policy in Britain during the war.
[24] See Ian A. Copley, *The Music of Peter Warlock* (London, 1979), 228.
[25] *MQ* 1 (1915), 381–6. Bartók's music is discussed on pp. 384–5.

these cobwebs from people's brains.' Reacting against Henry's analysis of Bartók's work in his *Egoist* article, Corder spoke his 'truth':

If, impressed by this and much more of the same sort of soulful utterance, the reader were so rash as to purchase any of Mr. Béla Bartók's compositions, he would find that they each and all consist of unmeaning bunches of notes, apparently representing the composer promenading the keyboard in his boots. Some can be played better with the elbows, others with the flat of the hand, none require fingers to perform nor ears to listen to. Yet you have to face the fact that audiences have sat, for the most part unmoved, while someone has gravely played the piano to them like a two-year old child. That men professing to be musical critics, while unable to compass the hypocrisy of pretending to comprehend this rubbish, have yet had the effrontery to encourage it, seems to me a shameful thing.

To Corder, Bartók's production was 'mere ordure'.

Other British writers were also pondering the directions taken by modern music at this time, although in a more balanced fashion. Leigh Henry, then a prisoner-of-war at the Ruhleben Internment Camp in Germany, delivered a series of lectures to his captive audience entitled 'Contemporary Composers'. Bartók's works were probably considered.[26] Several other commentators— Cyril Scott, Edwin Evans, A. Eaglefield Hull—also wrote during the war years about trends in contemporary music, but despite recognition of the significance of Schoenberg and Stravinsky, they made no mention of Bartók.[27] Only in a brief article, 'The Hungarian Note', in the February 1917 issue of the *Monthly Musical Record*, did D. C. Parker, a Glaswegian critic, speculate on the future of the progressive movement in Hungarian music:

Before the war a party committed to the dissipation of the confusion which has arisen through the juxtaposition of the Magyar and the gypsy was very active. It sought, I believe, to retrace the footprints of history and re-establish Hungarian music in its purity; it deplored the vicissitudes which the czardas has undergone at the hands of nomadic tribes. Both Béla Bartók and Zoltán Kodály, professors of the conservatoire of Buda-Pesth [*sic*], belong to this group. . . . Hungary is at a parting of the ways. Her musical salvation is possible only if she remains faithful to her own heritage—and the Iliad of her song reads like some highly-coloured page of Jókai— only if she persists in cultivating what seems 'wrong-headedness' if judged by the stereotyped standards of the conservatoire. By this means alone can she give us something which cannot be produced elsewhere.[28]

Despite wartime limitations, Philip Heseltine did manage to perform Bartók works publicly in Britain (as it then was). While in Ireland to avoid

[26] See *MS* 1 Jan.–4 Mar. 1916, for the introductory lectures in Henry's series.

[27] See, respectively: *MMR* 46 (1916), 134–5; *MT* 58 (1917), 347–51; *MO* Nov. 1917–June 1918.

[28] *MMR* 47 (1917), 33–4. Parker had established contact with Bartók before the war concerning Bartók's editorial work on Liszt's *Hungarian Rhapsodies*. See unpublished postcard, Bartók to Parker, n.d. [postmarked 22 May 1914], Department of Manuscripts, National Library of Scotland.

any chance of military service he took advantage of its current strife and
ventured to give a lecture-recital entitled 'What Music Is' in Dublin on 12
May 1918, which included in the programme a selection of short Bartók
piano works from *Bagatelles*, Op. 6, *Ten Easy Pieces* (1908), and *For
Children* (1908–9).[29] This was, however, the only technical breach in a
public silence which lasted for six years. The wartime embargo was, in
effect, only lifted for Bartók's works on 31 August 1920. As if resuming
where he had left off in 1914, Sir Henry Wood repeated the controversial
Suite No. 1, Op. 3 at a Promenade concert.[30] The music critic of *The Times*
considered the performance excellent, but found the suite disorganized and
lacking in development.[31] None the less, it had been 'well worth playing',
particularly because of Bartók's ingenious scoring. Alfred Kalisch wrote for
the *Musical Times* that the work had caused more vehement differences of
opinion than had any other modern work presented during the Promenade
season, but conjectured that it would find a permanent place in the
orchestral repertory. He clearly thought that the suite had been composed
recently: 'In his first stages he [Bartók] tried, with the fine recklessness of
youth, to begin where his most advanced predecessors left off; but it looks
as if now he were more cautiously retracing his steps and had nearly found
his proper place.' 'His proper place' was, one gathers, somewhere in the
mid-nineteenth century! In the recently established journal the *Sackbut*,
Cecil Gray briefly considered the performance, concisely concluding that
'the Béla beareth the bell away'.

  Wood's performance revived critical interest in Bartók. In October 1920
Leigh Henry featured him in the 'Contemporaries' series which he was
writing for the journal *Musical Opinion*.[32] Essentially a revision of his
Egoist writings of six years before, this article only looked at Bartók's pre-
war output, but did mention the triumphant Budapest première of his ballet
The Wooden Prince, Op. 13 in 1917. In the final paragraph Henry ascribed
a greater importance to Bartók's work than he had done before the war:

There is no single number of his compositions which I know of which does not at
once appeal to one's most evolved sensibility and intrigue one's intelligence. This
fact alone would render him worthy of close consideration: but beyond it Bartók has
given us a new interpretation of a tradition, and at the same time opened up to us a
new realm of musical possibilities. These achievements place him among the most
significant and important of contemporary composers.

[29] BL Add. MS 54197 (25 Apr., 5 and 14 May 1918). The background and reception to this
concert are considered in detail in Chapter 7, below.
[30] Years later Bartók recalled the significance of this first post-war British performance. See
BBE p. 521.
[31] Reviews mentioned: Anon., 'The Promenade Concerts', *The Times*, 1 Sept. 1920;
'London Concerts', *MT* 61 (1920), 689; 'Contingencies', *Sackbut*, 1 (1920–1), 224.
[32] *MO* 43 (1920–1), 53–4.

More authoritative was the article by the young critic Cecil Gray which appeared during the following month in the *Sackbut*, of which his friend Philip Heseltine was then editor.[33] Well informed, up-to-date, prophetic, and controversial, this article challenged British musicians to take Bartók's music more seriously.

Gray wrote in superlatives about Bartók's works of 1908–10, singling out the First String Quartet, Op. 7 for comparison with Beethoven's output in the genre: 'Bartók may have equalled his first string quartet; he has yet to surpass it. It is probably the finest achievement in quartet writing since Beethoven. This of course is sufficient reason to account for its neglect.'[34] Such evaluation could only provoke an English public that had never heard the work and, with a few exceptions, knew barely more than the name of this Hungarian composer. In more recent works, such as the *Two Pictures*, Op. 10 and the Second String Quartet, Op. 17, Gray found even greater affinities with the procedures of the older Beethoven, and concluded that these works 'take rank among the finest achievements of the century'.[35] A delight in paradox is evident in Gray's statement that 'Bartók is a great stylist precisely because he has no style', in justification of which he argued that Bartók's compositions revealed a mastery of all forms of expression while, at the same time, the composer's personality always emerged to create a highly individual utterance.[36] Although this opinion does not bear close scrutiny, it exerted a great influence on subsequent writers about Bartók, being repeated in one form or another by music commentators in Britain for several decades. When Bartók himself came to read this article he was most impressed by Gray's views on his significance in the musical world: 'In November I was the subject of a 12-page (approx.) article in a London music periodical in which the writer placed me in the ranks of the world's greatest composers, not merely the greatest living composers but of all time.'[37] History has supported Gray in this assessment, yet to many contemporary British music-lovers it undoubtedly seemed just another hare-brained opinion from one of the excessively intellectual young critics.

In November 1920 Heseltine initiated a correspondence with Bartók, which broke the Hungarian's almost total ignorance of British affairs over the past decade. Bartók was eager for such a contact. His self-imposed isolation in the last years before the war had been intensified during the years of combat, and prolonged by the military and ideological battles which ravaged Hungary in 1919. Only late in that year was he able to think of re-establishing professional links in Austria and Germany. Heseltine's unexpected approach presented the possibility of a more distant promotion

[33] 'Béla Bartók', *Sackbut*, 1 (1920–1), 301–12.
[34] Ibid. 303.   [35] Ibid. 304.   [36] Ibid. 305.
[37] Ibid. 306–8; letter, Bartók to János Buşiţia, 8 May 1921, in Hungarian, reproduced in English in *BBL* pp. 153–5.

of his music. For the public the first fruit of this contact was the appearance of Bartók's own article, 'The Relation of Folk Music to the Development of the Art Music of Our Time', in the *Sackbut*'s issue for June 1921.[38] Bartók spoke directly to Britain's musicians of his deep concern for the preservation of peasant music, and explained how features of this music permeated his own composition. With characteristic compression of thought he asserted the 'absolute artistic perfection' of peasant music, equal, only in miniature, to the perfection of the largest of traditional musical masterpieces.[39] He believed that a synthesis was possible between folk-music and the recent tendencies towards atonal writing, thereby suggesting the source of the intense dissonance found in many of his compositions of this time. To his mind it was in the harmonization of simple folk or quasi-folk melodies that the greatest potential for free, atonal use of all twelve chromatic degrees was offered.[40] Bartók also briefly considered the extent to which other composers drew on folk music: Stravinsky, considerably, but to monotonous effect; Schoenberg, not at all, with the result that his works were difficult to understand.[41]

By writing this article and encouraging a friendship with Heseltine and his colleagues, Bartók hoped to lay the foundation for a concert tour of Britain, during which he planned to present his more recent compositions.[42] His hopes were raised not only by the generous support of Heseltine and Gray, but also by renewed contacts with the three Arányi sisters—friends of Bartók since his Academy days—who were then living in London.[43] Of greater long-term significance to the promotion of Bartók's cause, however, was the help offered by the critic M.-D. Calvocoressi. Probably inspired by reading the recent *Sackbut* articles, he wrote of his own accord to Bartók. Since penning his *Musical Times* articles concerning Bartók in 1913 he had spoken about Bartók's music at a lecture-recital in Paris during early 1914, and had personally met the composer in that city shortly before the outbreak of war.[44] Bartók was naturally delighted that this influential critic, having since moved from Paris to London, was now freely offering his services. In a long letter of reply Bartók brought Calvocoressi up to date with his news: folk-music plans, mutual friends who had died, recent compositions by Kodály and himself.[45] He also expressed his fond hope of a tour:

I should very much like to go to London for a few weeks and to arrange some concerts of my works there. Undoubtedly it is a little difficult to bring this scheme

---

[38] *Sackbut*, 2 (1921–2), 5–11, reproduced in *BBE* pp. 320–30.
[39] *BBE* pp. 321–2.   [40] Ibid. 323–4.
[41] Ibid. 325–6.   [42] See *BBcl* pp. 30–11.
[43] The relations between Bartók and the Arányi sisters are considered in detail in Chapter 8, below.
[44] See M.-D. Calvocoressi, *Musicians Gallery* (London, 1933), 269; BBrCal pp. 199–202.
[45] Letter, Bartók to Calvocoressi, 31 July 1921, in French, in BBrCal pp. 202–5.

about. It will hardly be possible without reimbursement of my expenses, as my financial situation absolutely rules out a trip so far afield. I have two friends in London, Messrs. Philip Heseltine and Cecil Gray (both of them have come to see me this year, in Budapest). They also would be very pleased if I were to be successful in this scheme; but I do not believe that they can do anything for me in this matter.

Little by little, plans took shape. What worried Bartók most was that the receipts from such a tour would not cover his expenses. In a letter of 24 September to Calvocoressi he again stressed the necessity of covering all costs, and provided detailed estimates of transport, provision, and accommodation costs.[46] This nitpicking concern with finances was typical of Bartók throughout his life, but was certainly justified at this time. Hungary, in common with other defeated nations, was experiencing massive economic problems and an ever-growing rate of inflation—circumstances which made it impossible for the Bartóks even to live in a home of their own and ensured a frugal day-to-day existence.[47] Until venues and costs could be settled Bartók did not wish to delude himself with precise details of programmes or co-artists. During the final months of 1921 his fortunes, none the less, started to look up. He grew enthusiastic about performing with Jelly, the youngest of the Arányi sisters, and willingly complied with her request for a violin sonata (the First).[48] The Arányi sisters, accordingly, became more dedicated to helping with his proposed tour. On the Continent, his opera *Duke Bluebeard's Castle*, Op. 11 and ballet *The Wooden Prince*, Op. 13 were accepted for performance early the following year in Frankfurt-am-Main, and an invitation was issued for him to perform in Paris. 1922 was promising to be the year for Bartók's long-overdue elevation to a truly international status.

By December 1921 the Arányis had made arrangements for two British engagements, at Aberystwyth in Wales and Hammersmith in west London.[49] They suggested, as well, an evening of violin and piano sonatas, but pointed out that a profit could not be guaranteed as it depended on the sale of tickets. With promised fees of £30, augmented by the unexpected arrival of a fee of nearly £5 for his *Sackbut* article, Bartók considered that the trip was financially viable, and proposed alternative dates for it in mid-March and mid-April.[50] Further encouraging news came from England on 22 December, when the influential critic and founder of the British Music Society, A. Eaglefield Hull, wrote to assure Bartók of his full support for the planned tour.[51]

Even so, Bartók was still fretting over the costs, panic-stricken that by

[46] BBrCal pp. 205–7.
[47] As an indication of the Hungarian rate of inflation at this time, one gold crown was worth 9.9 paper crowns in August 1919, 50.7 paper crowns in June 1921, and 18,400 paper crowns by May 1924. See I. T. Berend and G. Ránki, *Hungary: A Century of Economic Development* (Newton Abbot, 1974), 99–110. [48] BBrCal pp. 207–9.
[49] BBrCal pp. 209–10.   [50] Ibid.   [51] MTA BA-B 3477/90.

staying in Britain too long he might run into debt. On 2 February 1922 he wrote to Calvocoressi: '. . . I can spend only a short time since the different honoraria are not sufficient for a longer stay. . . . Actually, it is a pity that I cannot also play publicly in London. If it were not for the opportunity of playing in Paris it would not be worth travelling to England for these two private concerts.'[52] Finally, at the last possible moment, on 17 February, Bartók made a definite commitment to go to Britain during the following month, but only for two weeks.[53] As it turned out, he spent over three weeks there, played in many more concerts than expected, and covered costs many times over![54]

While these plans for a tour were being refined, Bartók and his music were becoming fashionable topics in British musical circles. True to his word, Calvocoressi had been promoting Bartók's cause frequently in his journal articles and reviews. Through his column 'Music in the Foreign Press' in the *Musical Times*, he kept readers abreast of Bartók's latest activities and achievements.[55] More on the offensive in the *Monthly Musical Record*, he took sneering critics and intolerant members of the public to task for their glib dismissal of Bartók's and Stravinsky's works.[56] Others, too, played their part in giving Bartók a higher profile. In October 1921 the *Musical Times* published an interim review of his *Three Studies*, Op. 18, composed in 1918:

These works give us almost the first glimpse of composers' activities in Germany and Austria since 1914. Those of us who anticipated a return to simplicity will not find it here. . . . It is impossible to realise mentally the effect of the Bartók Studies, and, as a mere reviewer can do no more than labour painfully through them at the keyboard, it would not be fair to express an opinion. Bartók, we know, is a composer who counts, and we may presume that in this case he says something worth saying. What that something is we shall know when one of our pianoforte recitalists gives us a chance of hearing it.[57]

In the November 1921 issue of the *Monthly Musical Record* Edwin Evans, the chief force behind the reformation of contemporary music concerts after the war, gave formal notice of Bartók's plans to visit Britain.[58] Under the title 'London, the World's Music Centre', Evans surveyed the upsurge of interest in modern music over the previous decade, and listed the great progressive composers who had visited Britain in that time: Skriabin, Schoenberg, Strauss, Reger, Debussy, and Ravel. 'The exposition of the new movement up to date will not be complete', he continued, 'until occasion is

[52] Letter, Bartók to Calvocoressi, 2 Feb. 1922, in German, in BBrCal pp. 211–12.
[53] BBrCal p. 212.   [54] See *BBcl* p. 329.
[55] See *MT* 62 (1921), 334–5, 625–6, 839.
[56] 'Does To-day's Music Bewilder the Public?', *MMR* 51 (1921), 174–5.
[57] H. G., 'Pianoforte Music', *MT* 62 (1921), 701.
[58] *MMR* 51 (1921), 241.

made for the appearance of that great musical creator, Béla Bartók. It is good news to hear that he is desirous of visiting London.'

Bartók himself contributed further to public awareness of his activities by writing an article entitled 'The Development of Art Music in Hungary', which appeared in the *Chesterian*'s issue for January 1922.[59] This article provided background information which was most necessary for a proper understanding of national influences in his music: nineteenth-century attempts at creating distinctively Hungarian music, and the effect of folk-music upon composition in this century. But then, with characteristic self-effacement, Bartók went on to consider the musicians who had contributed most to this folk-movement—Kodály and Lajtha—and those who had stood aloof from it yet still produced works of repute—Dohnányi, Szántó, and Weiner. Nowhere was Bartók himself mentioned. This omission was pointed out in a footnote appended by the journal's editor, Georges Jean-Aubry: 'The Editor cannot refrain, with all due to the high achievement of Kodály, from drawing attention to the omission, due to the Author's modesty, of a still more important name.'[60]

In the concert-hall, too, Bartók was being heard. At the Queen's Hall on 15 October Sir Henry Wood realized one of his intentions for the 1914 season with a performance of Bartók's *Rhapsody*, Op. 1 in its version for piano and orchestra. The soloist, Auriol Jones, was the same as originally cast.[61] Around this time Bartók's *Two Portraits*, Op. 5 for orchestra was also performed by Sir Henry, probably at the smaller venue of the Aeolian Hall.[62] A concert performance of the music to *The Wooden Prince* was also mooted, but it did not materialize.[63]

Early in 1922 both Calvocoressi and Heseltine wrote extensively about Bartók to help raise further the level of interest in his visit. Calvocoressi summarized the state of contemporary Hungarian composition in the *Monthly Musical Record* of February 1922, simply distinguishing a Right Wing, little influenced by folk-music, from Bartók's supporters on a Left Wing.[64] The following month he turned his journalistic skills to providing an uncomplicated introduction to Bartók's music, based mainly on his pieces for children.[65] There, Calvocoressi explained why he advocated so strongly the study of these easy pieces:

---

[59] *Chesterian*, NS no. 20 (Jan. 1922), 101–7. [60] Ibid. 104.

[61] This concert appears to have passed unreviewed.

[62] See *A Dictionary of Modern Music and Musicians*, ed. A. Eaglefield Hull (London, 1924), p. 30. The author has been unable to ascertain the exact date of this performance. The dictionary reference may possibly result from a confusion with a performance of the work in early 1923. The date of 11 May 1914 given by Kenneth Thompson (*A Dictionary of Twentieth-century Composers (1911–1971)* (London, 1973), 'Bartók' entry) for Wood's first performance of this work would appear to be incorrect.

[63] Cf. *MT* 62 (1921), 787–9 and *MT* 63 (1922), 40.

[64] 'Hungarian Music of To-day', *MMR* 52 (1922), 30–2.

[65] 'Béla Bartók: An Introduction', *MMR* 52 (1922), 54–6.

as pure music, they are delightful; for educational purposes, they are invaluable. I know no music better suited for broadening the beginner's outlook whilst providing excellent and varied practice in reading, phrasing, colour, and style. They are, naturally, the best introduction to Bartók's music, which in turn is, in my opinion, the best with which to start the study of recent developments in music generally.

Heseltine's article, 'Modern Hungarian Composers', which appeared in the *Musical Times* on 1 March—just ten days before Bartók's arrival in Britain—was ideally timed to influence public opinion.[66] He identified the two string quartets as Bartók's best works, as they showed his 'singular genius' most clearly, and concurred with Gray in claiming these quartets as the most significant such works since Beethoven.[67] Perhaps foreseeing the lines of attack of London's more conservative critics during the approaching tour, Heseltine was at pains to distance Bartók's work from that of better-known contemporaries such as Schoenberg and Stravinsky:

Bartók is a 'modern' whose originality owes nothing to sensationalism, eccentricity, or 'revolutionary' ideas, and does not depend for its recognition upon the postulation of a world from which the great masters of the past are rigidly excluded. He is, moreover, singularly free from the influence of other contemporary composers; and those who make the acquaintance of his work in 1922 and observe therein that simplicity of texture, directness of expression, and freedom from conventional forms and formulae . . . should bear in mind that much of Bartók's best and most characteristic work is already fourteen years old.[68]

The concluding paragraph of the article emphasized the importance of Bartók's visit. His music was little better known in his homeland than in London, Heseltine stated, and a true recognition of his talent in as important a musical centre as London would greatly assist such acknowledgement in Hungary. Not to miss out on a last chance of publicity, Calvocoressi penned two further short articles, which appeared on 11 March just after Bartók had arrived in the country.[69] Midway through one of these, in the *Musical News and Herald*, Calvocoressi took stock of where Bartók's music then stood in musical life:

It would have been natural to expect that the beauty and originality of his output would have ensured its speedy diffusion. Such has not been the case so far; his music is making headway, indeed, but slowly, and with the same unobtrusiveness which marks Bartók's doings and everything connected with him. Some of his works are known in every country; but few, if any, have reached the public at large. Music lovers read them, play them, talk of them, and long to hear them; but the concert repertory shows few signs of their existence. Not even the two admirable String Quartets nor the principal pianoforte pieces have yet come to their own on the platform.

---

[66] *MT* 63 (1922), 164–7.   [67] Ibid. 165.   [68] Ibid. 167.
[69] 'Béla Bartók', *MNH* 11 Mar. 1922, p. 306; 'Music of the Day: Béla Bartók', *DT* 11 Mar. 1922.

In spite of the performances mounted by Sir Henry Wood, at the time of Bartók's visit enthusiasts in Britain had heard nothing of his orchestral writing after 1908. His stage works remained a mystery, as did all his post-war compositions: their existence known, their reality unexperienced. Of the London critics who eagerly awaited him only Calvocoressi, Heseltine, and Gray knew him personally, but they had never reviewed his performances. Ernest Newman had indeed reviewed Bartók's Manchester performance of 1905, although he did not recall having done so.[70] By March 1922, none the less, the musical public had repeatedly been alerted to Bartók's significance. He was recognized as a composer ascendant on the international scene, but still more a name, like Schoenberg, representing an obscure school of ultra-progressive composition beyond the Channel, than a flesh-and-blood musician associated with a wide variety of real compositions. During his short stay an unexpectedly wide variety of his compositions would be presented, to the shock of many in Bartók's audience.

[70] See 'The Week in Music', MG 30 Mar. 1922.

3.

1922: IN THE LIMELIGHT

In 1922 Britain afforded Bartók a reception quite beyond his expectations. The caution with which his visit had been planned had resulted from a vast underestimation, by the composer himself, of the British public's interest in his music. Through the many activities of his London friends over recent months, 'Bartók' had become a well-known name. Widespread curiosity had been aroused. The critics, in particular, flocked to hear him in their dozens. For many of them, however, Bartók's most recent works would only inspire a bewildered response. These conservative critics would find that in harmonic dissonance, fragmentation of melodies, and apparent disintegration of musical structure, these works went far beyond the level of modernity which they had expected, and had come ready—if need be—to ridicule. The very point of music was now brought into question, and until Bartók's techniques and philosophies were better understood many of the critics felt that he needed to be given some benefit of the doubt. The sincerity with which he presented his music, as well as the zeal and loyalty of his small body of supporters, also encouraged many commentators to hold back from expressing their innermost worries. This was a period of 'honeymoon' for Bartók, when tentative approval or complete withholding of judgement were the most common responses. In later years Bartók's relationship with the British critics would be put to a more real test, with none of the politeness of a first encounter, but in the meantime their initial interest and general lack of hostility would have been interpreted across Europe as the first sign of wider international recognition. Just as the 1917 Budapest performance of the ballet *The Wooden Prince* had catapulted Bartók into a more deserved position in Hungary's musical life, so too did the 1922 visit to Britain gain for him a more secure place in the front rank of European composers.

Bartók's expression of disappointment at the small number and private nature of the engagements arranged for him had stung his helpers in Britain into action.[1] They prevailed upon their friends and musical associates to provide more venues, especially public ones, at which Bartók could perform. In the end, the tour consisted of three public (or open association) concerts—in London, Aberystwyth, and Liverpool—and probably four private London engagements. Private concerts had traditionally been an integral part of the musical life of London's 'well-to-do', although since the war, with the straitened circumstances of many patrons, the private system

[1] BBrCal pp. 211–2.

had started to show signs of decline.[2] Coming from a country where private concerts were by then seldom held, Bartók was pleased to have access to this form of music-making, especially in light of the difficulties he had experienced in arranging any London concerts during his pre-war visits. Despite the private nature of these concerts a considerable charge was often imposed on guests to help cover the fixed fee already negotiated with the performer. The guaranteed nature of the fee was appealing to a musician. Bartók, for instance, received fees on this tour of between £10 and £30 for his private engagements, depending on the patron, venue, and whether or not tickets were sold.[3] Public concerts, although giving more people access to the performer, and thereby propagating the works presented more widely, frequently involved an element of financial risk. With a packed hall of paying guests the performer could do handsomely, but generally, in an oversubscribed market, earnings from single public concerts were slender. For many European artists, with weak home currencies, it was a dangerous gamble to make the large outlay for hiring a London concert-hall. Even British musicians were having problems, which were magnified if their programmes included contemporary items.[4] In view of this situation Bartók's extreme caution in committing himself to the tour was certainly not misplaced.

From the moment of his arrival in London on 10 March Bartók was 'news'. Brief reports appeared in several newspapers,[5] a prelude to more intensive press coverage and requests for interviews from publications as varied as the *Pall Mall Gazette* and the *Christian Science Monitor*.[6] Bartók's name only started to appear in the headlines, however, after the first concert, held on 14 March. This private recital had been arranged by the Arányis at 18 Hyde Park Terrace, the residence of the Hungarian chargé d'affaires. After some hesitation, Bartók and Jelly Arányi decided to introduce his (First) Violin Sonata (1921)[7] as the central feature of the programme. The remainder of the programme consisted of six Bartók piano pieces, and violin works by Spohr and Beethoven.[8]

As it had been a private concert, Bartók was surprised at the extent to which reviews appeared in the newspapers—but pleasantly so.[9] *The Times*

---

[2] For a more detailed description of London musical life before the First World War see Malcolm Gillies, 'Grainger in London: A Performing History', *Musicology Australia*, 8 (1985), 14–23.

[3] *BBL* p. 157. Bartók was surprised at the high entrance charge of 10s. 6d. to the first of his private concerts. See *ZT*.vii pp. 234–7.   [4] See Editorial, *MO* 45 (1921–2), 849.

[5] *BBL* pp. 157–8.   [6] Ibid., and MTA BA–B 3477171, respectively.

[7] Mary Dickenson-Auner and Eduard Steuermann had premièred the sonata in Vienna on 8 February. This performance was sanctioned by Bartók—surprisingly, in view of his whole-hearted dedication of the work to Jelly Arányi.

[8] MTA BH 2049/93 (preliminary); MTA BH 2049/94 (final).

[9] Reviews mentioned: Anon., 'Hungarian Music', *The Times*, 15 Mar. 1922; H. C. C., 'Béla Bartók', *The Times*, 18 Mar. 1922; R. C., 'Béla Bartók', *DM* 16 Mar. 1922; M.-D.

even reviewed the concert twice. In its first critique, appearing on the following morning, the unnamed writer naturally concentrated on Bartók's own works in the programme, admiring them, but carefully avoiding any precise evaluation: 'It is a curious feeling to listen, without any sense of connexion, to these strange passages—cries, it might be, of suffering men or wounded animals, sprung from the soil, perfectly sincere. We get restless and excited, and feel at last a sort of triumph in having escaped from the conventions of Nature.' In closing, the reviewer acknowledged the 'extraordinary privilege allowed to the select audience'. On the following Saturday the new sonata was the subject of a longer article by the chief music critic of *The Times*, H. C. Colles. He called for listener 'dissociation'—suspension of all previously held musical values—to allow Bartók's work to establish its own procedures and be appreciated on its own terms. Colles then provided a brief analysis of the work. Although only an impression after one hearing, this analytical sketch is notable for highlighting the conflict between the violin and the piano as they pursue their different paths, and also for pointing out how little the work owes to traditional procedures of repetition of material. Given these innovations, the continual sense of forward direction in the work had, in Colles's opinion, to be attributed to 'some manifestation of the rhythmic impulse'.

More colourful, but still tentative in its conclusions, was Richard Capell's review in the *Daily Mail*:

It was with him [Bartók] as with Mr. Igor Stravinsky—the contrast was remarkable between a slight, small man of most modest bearing and an art of hair-raising ferocity. The piano pieces did not give us his full measure. They were pleasant, fresh, exotic enough, the music of a sort of Hungarian Grieg. The remarkable sonata was a more powerful affair. This Bartók is bent on jilting us out of any dull, smooth paths. It is not by any means all brute force (which can be dull as anything), but has an edge, sharpened by every new-fangled ingenuity, and its way will be made easier here since we know so well this music's chief embodiment, Stravinsky's 'Rites of Spring' [*sic*].

The concert was reviewed further afield than London. The correspondent for the *Glasgow Herald*, none other than M.-D. Calvocoressi, was also wary of making definitive statements, although he certainly knew better than to attribute too much of Bartók's technique to Stravinsky. He examined, first of all, Bartók's pianism, praising his 'perfect sense of rhythm', his 'rare gift for incisive, bold phrasing', and asserting Bartók's right to stand in the front rank of musical interpreters. Jelly Arányi's playing in Bartók's sonata also elicited the highest praise, with compliments on her musical understanding, technique, and sheer energy. But Calvocoressi shied away from any judgement of Bartók's new composition. It was 'direct and consistent

Calvocoressi, 'The Week's Music in London', *Glasgow Herald*, 22 Mar. 1922; Anon., 'Színház és Művészet', *Pesti Napló*, 23 Mar. 1922.

1. Bartók and his sister Elza, 1904

Oxford Street, Manchester

Valentines Series

2. Bartók's postcard to his piano-teacher István Thomán. February 1904

3. Bartók with Jelly Arányi (left) and Adila Fachiri (née Arányi). London, March 1922. This photograph originally appeared in the *Daily Sketch*, 24 March 1922

PROGRAMME.

(1). Sonata for Violin and Pianoforte in E major - Bach.

Adagio—Allegro—Adagio ma mon ianto—Allegro.

Jelly d'Arányi and Béla Bartók.

(2). Pianoforte Solos.

(a). For Children
(b). Sonatine
(c). Burlesque No. 2 ("Un peu gris") (from Op. 8c)
(d). Evening in the Country (from "Ten Easy Pieces")
(e). Allegro Barbaro
(f). Bear Dance (from "Ten Easy Pieces")
(g). Roumanian Dance No. 1 (from Op. 8a)

} Bartók.

Béla Bartók.

(3). Sonata for Pianoforte and Violin in A major, Op. 47 } Beethoven.

Adagio sostenuto, Presto—Andante con Variazioni—Presto.

Béla Bartók and Jelly d'Arányi.

7, Sydney Place, S.W.7.
31st March, 1922.

THE ABBEY SCHOOL, MALVERN WELLS

Pianoforte Recital

BY

BÉLA BARTÓK

Friday, May 4th, 1923

I. SCARLATTI - (a) Three Piano Pieces
   BARTÓK - (b) From "15 Hungarian Peasant-Songs"
            (c) No. 6 (Tema con variazioni)
            (d) Nos. 7—15 (Old Dance Tunes)

II. BARTÓK - (a) Second Elegy (from Op. 8b)
             (b) Bear Dance
             (c) Allegro barbaro
             (d) Evening in the Country
             (e) Three Burlesques
                 (A Quarrel—un peu gris—molto capriccioso)

III. DEBUSSY - (a) Mouvement from "Images"
               (b) Danseuses de Delphes
               (c) General Lavine eccentric
   from "Préludes" (d) Les fées sont d'exquises danseuses
               (e) Ce qu' a vu le vent d'Ouest
               (f) Les collines d'Anacapri

IV. BARTÓK - (a) Roumanian Peasant-Dances
             (b) First Dirge
             (c) Sonatina
             (d) First Roumanian Dance

4. Programme for a private concert at the London home of Duncan and Freda Wilson. March 1922

5. Programme for a recital at a girls' school in Malvern Wells. May 1923

6 & 7. Two views of Normanhurst Court, Battle, a girls' school where Bartók performed in
May 1923. These photographs were taken some years before Bartók's visit

8. Letter from Erik Chisholm to Bartók concerning arrangements for his first concert in Glasgow. October 1931. On the left-hand side, Bartók has worked out the details of his programme, including timings

HUNGARIAN COMPOSER ARRIVES IN GLASGOW

Bela Bartok, the Hungarian composer (third left), photographed with Sir W. Burrell, Hungarian Consul (right), and Lady Burrell on his arrival in Glasgow last night. He will appear at a concert of his music in the Stevenson Hall, Glasgow, this evening.—" Bulletin " Photograph.

9. From the *Bulletin and Scots Pictorial*, 29 February 1932

FAMOUS COMPOSER.—Bela Bartok, the famous Hungarian composer and pianist, being greeted by Mr Erik Chisholm and Mr Ernest J. Boden, of the Active Society for the Propagation of Contemporary Music, on his arrival in Glasgow last night. He gives a concert in St. Andrew's Hall to-night.

10. From the *Evening News* (Glasgow), 2 November 1933

11. Bartók and Mária Basilides await instructions from the announcer in a BBC broadcast from Dohnányi's home in Budapest. September 1935

12. Bartók in London. January 1936

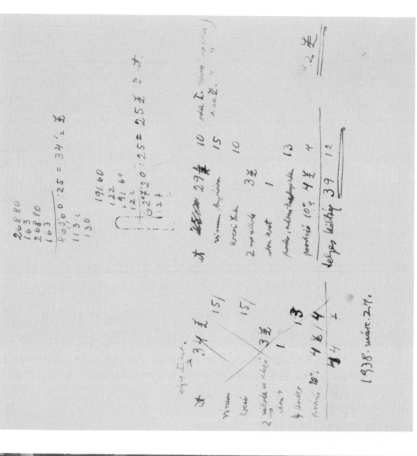

14. Bartók's estimates for his trip to Britain in June 1938, when he was accompanied by his wife. His final decision was to travel first class out and second class back

13. On tour with his second wife, Ditta. 1938

throughout', he claimed, but demanded another hearing. Even further afield, in Budapest, the *Pesti Napló* reported on 'Béla Bartók's Successes in London' and hailed him as 'the pride of modern Hungarian music'. Referring to the sub-title of Colles's review, 'A Foreign Language', the Hungarian article concluded: 'Bartók's music seems a foreign language to the English music critic, but it is that type of language which interests foreigners.'

Encouraged by this warm reception in London, Bartók took the train on the following day to Aberystwyth, in Wales, for his first public performance. This occurred on Thursday 16 March as part of the season of weekly concerts organized by the music department of the local University College of Wales. Bartók had been invited to perform there through an approach of the Arányi sisters to their friend Walford Davies, then professor at the College and soon to gain national fame as a pioneer of music commentaries on radio. So popular had these weekly concerts become that they had recently been moved to a parish hall so that extra seating could be provided.[10] In this small coastal town Bartók found time to relax after the hectic days in London, and wrote the first letters home telling his good news. In the longest one, to his mother, he described his reception in London and outlined his coming engagements, providing a full list of fees and including the information that the E. L. Robinson concert agency had agreed to bear the risk for the one public London concert. With schoolboyish glee he continued: '. . . as I am staying here as the guest of a very friendly couple (I haven't been spending money on anything at all so far), and expect to receive a total of 1,500 francs in Paris, it looks as if I shall be able to bring lots of money home with me'.[11] To his sister Elza he did not go into such details, merely sending her a postcard of 'Aberystwyth from Constitution Hill' with some general news: 'It is a magnificent, hilly region, with a wild, rocky seashore, and there isn't a single cloud in the sky. I went for a nice walk with—a Hungarian man who is a mathematics lecturer at the university here! The concert is this evening.'[12]

The programme to which Bartók contributed was a mixed bag.[13] He played two groups of his own piano pieces, including the *Allegro barbaro* (1911) and the *Suite*, Op. 14, and took part in Beethoven's Piano Trio in E flat major, Op. 70 No. 2 with local artists. Two Parry songs and well-known Handel and Bach works rounded out the programme. In the company of such old favourites Bartók's music bewildered the audience, by all reports. Walford Davies himself, grandiloquent in his introduction,

[10] See MO 45 (1921–2), 505. The admission arrangements were enlightened for the time, particularly in allowing free entry to university students.
[11] Letter, 16 Mar. 1922, in Hungarian, reproduced in English in *BBL* pp. 157–8. By 'here' Bartók meant London, as he was staying in a hotel in Aberystwyth. The 'friendly couple' was probably Duncan and Freda Wilson, although it may have been Robert and Dorothy Mayer.
[12] Postcard, 16 Mar. 1922, *BBcl* p. 327.   [13] MTA BA-B 2049/90.

could afterwards only comment, 'Baffling, isn't it?'[14] Local newspaper and journal correspondents could do little better in evaluating this strange music. In the *Welsh Gazette* the critic recorded the audience's warm appreciation of Bartók's participation and looked on his appearance as a fitting climax to the 'brilliant list of sessional musical events' arranged by the College.[15] Readers of the *Cambrian News and Welsh Farmers Gazette* read of a 'masterpiece of technique which gripped the attention of the vast audience. . . . It made the strongest appeal to the intellect and, though its meaning and message was somewhat obscured to the uninitiated, it gave a glimpse of hitherto undreamt of possibilities.'[16] Many reviewers were, therefore, all too happy to be side-tracked into lengthy descriptions of a local squabble concerning publicity and the kudos accruing from the event. The *Western Mail*, a Cardiff newspaper, started the debate:

Music-lovers at the collegiate town who realise the significance of this outstanding event are asking why no notice of it has been published: but the more pertinent question is: Why was not notification given to the press? Here is an eminent foreign musician whom some of the critics rank amongst the foremost composers of the day. . . . Bartók and his music have been so much discussed that all musical Great Britain has been eagerly awaiting his forthcoming first visit to London. Yet Aberystwyth forestalls London, and no one is told anything about it, and the mere fact only leaks out by accident! Why this secrecy?[17]

Replies from the musical fraternity of Aberystwyth revealed some smugness. Walford Davies answered simply: 'We are not in the habit of advertising our doings, and do not see why we should. We told all our members.'[18] But for 'Cymro', the Aberystwyth correspondent of *Musical Opinion*, the opportunity to score points was irresistible:

The fact that the Aberystwyth visit forestalled that of London and Liverpool would have probably passed unnoticed were it not that the critic of a Cardiff daily felt compelled to wail 'Why wasn't I told? Why was the press ignored?' The whole truth consists in the fact that, although the hall in which these concerts are held is of a good size, accommodation is taxed to the utmost, and the appearance of artists of the widest repute is a frequent occurrence.[19]

Bartók's movements after this concert are not completely clear. From a copy of a telegram left behind by him at Aberystwyth, however, it is clear that he spent a day at the home of Philip Heseltine, near Newtown, Powys, before returning to London, perhaps for a private concert on the evening of 18 March in Hammersmith.[20] Although the securing of this engagement is

[14] Ian Parrott, 'Warlock in Wales', *MT* 105 (1964), 741.
[15] Anon., 'Aberystwyth', *Welsh Gazette*, 23 Mar. 1922.
[16] Anon., 'Béla Bartóck [sic]', *Cambrian News*, 24 Mar. 1922.
[17] *Western Mail*, 20 Mar. 1922. See also *MNH* 25 Mar. 1922, p. 377.
[18] *Western Mail*, 22 Mar. 1922.
[19] 'Music in Wales', *MO* 45 (1921–2), 776.   [20] MTA BA-B 3477/86.

mentioned in one of Bartók's letters to Calvocoressi, no further details are known.[21]

On the following day, 19 March, Bartók was the guest of honour at a reception given by the singer Dorothy Moulton, who was the wife of the industrialist Robert Mayer. As editor of the *Musical News and Herald*, Edwin Evans attended this reception. He found Bartók socially a shy man, with a total absence of pretension: 'somehow or other, he contrived to get continually mislaid, involving a careful search each time he was wanted. Naturally this endeared him to everyone present.'[22] Writing to his mother the next day, Bartók reported: 'I meet an awful lot of people, so that I am getting quite confused. Last night, I went to a frightfully "distinguished" party (i.e. all musicians and critics) given by a rich woman, a singer.'[23] Being on his first visit to an English-speaking country in seventeen years he naturally found the language a problem.[24] After the reception Bartók presented a programme chiefly of his own piano music—including the *Suite*, Op. 14 and his recent *Improvisations*, Op. 20—although he did accompany his hostess in a short collection of Hungarian folksongs. In his review Evans paid close attention to the qualities of Bartók's piano-playing:

It is this [percussive] quality of the piano that permits the abrupt, uninflected juxtaposition of sonorities which plays a vital part in contemporary music. In his playing Bartók gives an even more convincing illustration of this than I had anticipated. Though he gives full value to the possibilities of inflection as occasion demands, he is quite uncompromising in the crisp vigour of his touch when the music assumes that character.

Evans only passed the most general comments about the music itself, although he did recognise the 'intensely individualistic' quality of Bartók's art. Forty years later, in a short article for the children's magazine *Crescendo*, Dorothy Moulton looked back on this or a similar event:

He [Bartók] never apologised for himself, he was incapable of writing a note he did not believe in, he was very reserved and very proud and all this did not help him to win the favour of the musical public in Budapest. This mattered less, because he was acknowledged abroad as one of the greatest of contemporary composers. He came to London in 1923 [*sic*], and I had the honour of having him in my house as guest; he played at a party to which all the pianists in London came—and though one of them

---

[21] BBrCal pp. 209–10.

[22] 'In London', *MNH* 25 Mar. 1922, pp. 372–3, partially reproduced in 'Bartók als Pianist in London', *Musikblätter des Anbruch* (Vienna) 4 (1921–2), 155.

[23] Letter, in Hungarian, reproduced in English in *BBL* p. 158.

[24] Bartók had taken English lessons while studying at the Budapest Academy of Music (*BBcl* pp. 76–7), and took some further lessons before his visit to Britain in 1905 (*BBcl* p. 143). In several letters to foreign correspondents during the early 1920s he stated his good reading-knowledge of English, but acknowledged a lack of confidence in other uses of the language (BBrCal pp. 207–9, *BBlev*.v p. 270).

afterwards said 'I wonder your piano could resist such treatment': he had a tremendously hard and percussive touch—they all realised that he was a master.[25]

The level of interest in his tour still amazed Bartók. His public concert on 24 March was announced as one of the week's chief concerts in *The Times*, and later Bartók's photograph appeared in the same paper.[26] He was flattered by the fact of its appearance, rather than its likeness: 'it's certainly awful enough', he later wrote to his mother.[27] Socially, too, Bartók was a popular figure. Georges Jean-Aubry, editor of the *Chesterian*, held a luncheon for him on Monday 20 March at which L. Dunton Green, London correspondent of *La Revue musicale*, was also present.[28] Bartók noted: 'Today I lunched with some French people. I have to speak first French, then English (sometimes German); I falter always as best I can, but I get all mixed up by having to keep changing from one language to another.'[29] The following week he was again invited to meet these critics, and also the composer Lord Berners, at a small dinner-party in Holland Park.[30] Letters of support and interest in his activities began to arrive. On 22 March, perhaps as a result of meeting Bartók at the Moulton reception, Leigh Henry wrote to him giving details of his early articles. He concluded: 'In any possible way always count on my admiration, sympathy and utmost support.'[31] Sir Hugh Allen, the Director of the Royal College of Music, wrote on the same day to Adila Fachiri (née Arányi), hoping to arrange a meeting with Bartók and inviting him to play before the students for a fee of 5 guineas.[32] Whether deterred by the small fee or because of other appointments, Bartók did not accept this invitation. When, however, another last-minute offer of a performance arrived, this time from the British Music Society in Liverpool, he did decide to accept it.[33]

So far, Bartók had been heard by only a fraction of London's musical community. His Aeolian Hall concert on 24 March was, therefore, very well attended. The battery of over twenty critics included such celebrities as Ernest Newman, Percy Scholes, Eric Blom, and Edwin Evans. Both the quality and the popular press were represented. Some of the more senior critics penned reviews in more than one publication, helping to make this the most intensely scrutinized of any of Bartók's nearly fifty performances in Britain. This great interest was, however, partly a product of the time.

[25] 'Glimpses into Lives of Composers: Bartók', *Crescendo*, no. 114 (Oct. 1961), 8–12.
[26] 'This Week's Music', *The Times*, 20 Mar. 1922; photograph, *The Times*, 22 Mar. 1922.
[27] Letter, 5 Apr. 1922, *BBcl* pp. 329–30.
[28] See letter, Georges Jean-Aubry to Bartók, 15 Mar. 1922, in French, *DB*.iii p. 114.
[29] Letter, Bartók to his mother, 20 Mar. 1922, in Hungarian, reproduced in English in *BBL* p. 158.
[30] MTA BA-B 3477/10. Among Bartók's scores are four pieces by Gerald Tyrwhitt (Lord Berners), copied in Bartók's own hand (MTA BH I. 52). Evidence of several different kinds suggests that these were copied by Bartók during this visit to Britain. See *DB*.v p. 166.
[31] MTA BA-B 3477/126.   [32] MTA BA-B 3477/8.
[33] MTA BA-B 3477/52.

Despite ever-present financial problems, London was then experiencing a peak of involvement in contemporary music. As Edwin Evans explained later in the decade:

We were then hearing more modern music than any capital in Europe, and I believe, any city in America. . . . Today we hear very little that is new unless we go abroad in search of it; we have lost contact with the currents of musical opinion at home and the conservatives hold practically undisputed sway. . . . in London, the unprecedented activity of those post-war seasons has left an aftermath of musical indigestion. Perhaps we overdid it.[34]

Jelly Arányi and singer Grace Crawford collaborated with Bartók in the programme, which featured works by Mozart and Kodály as well as Bartók. As with the programme for the concert on 14 March, the performers had second thoughts about the placement of items—Bartók's sonata in particular; they eventually presented the sequence given below.[35]

I. Violin Sonata in D major, K. 306   Mozart
JELLY ARÁNYI and BÉLA BARTÓK

II. a. Suite Op. 14   Bartók
   b. Improvisations Op. 20   Bartók
BÉLA BARTÓK

III. Four songs from Eight Hungarian Folksongs (1907–17) Bartók
   1. Gott, ach Gott im Himmel
   2. Wege schüttet man im Walde
   3. Wollt ich in die blauen Berge suchen gehn
   4. Frauen, Frauen lasst mich Euch Genossin heissen
GRACE CRAWFORD and BÉLA BARTÓK (accompanist)

INTERVAL

IV. [First] Violin Sonata (1921)   Bartók
JELLY ARÁNYI and BÉLA BARTÓK

V. a. Two Piano Pieces   Kodály
   Lento from Op. 3—Epitaphe from Op. 11
  b. Burlesque, from Op. 8c   Bartók
  c. Romanian Dance, from Op. 8a   Bartók
BÉLA BARTÓK

The reviews of the concert provide an excellent study in the prevailing diversity of British music criticism, in both matter and manner. While some critics attempted to give a balanced account of all items in the programme

---

[34] 'Half-Time in England', *Modern Music*, 3. 4 (May. 1926), 13–15.
[35] MTA BH 2049/92 (preliminary), MTA BH 2049/91 (final).

and to consider performing as well as compositional aspects, the majority directed their attention steadfastly to the most significant work presented, the new Violin Sonata. To a man, they experienced difficulty in discerning some meaning in that work: a few sought to account systematically for the ways in which Bartók dealt with the various elements of music, but most resorted to superficial 'blow-by-blow' accounts of its most obvious features. Indeed, the popular press, perhaps mindful of its readers, paid almost no attention to what was played, preferring to focus on issues of personality and sensation.

The occasional general comments on Bartók as a pianist were so varied as to make one question whether the same performance was being reviewed. Percy Scholes, in a cursory review in the *Observer*, was the most negative:

This great and much-misunderstood composer gave a recital on Friday, and—I am one of those who misunderstand him! What the Editor of 'Musical News' [Edwin Evans] describes in his yesterday's issue as 'the abrupt, uninflected juxtaposition of sonorities' (blessed words!) is to me a hard, cold rattle of a keyboard violently attacked in chance combinations of keys and notes, with the stiffened metal muscles of a jerkily rhythmic automaton. Pray for me, E. E.[36]

Ernest Newman, in the *Manchester Guardian*, also criticized Bartók's pianism, but more soberly. He focused on his stage presence:

He is altogether too modest. His platform manners are not of the sort to make an impression on any audience. He comes on and goes off almost apologetically; he does not seem to have studied even the rudiments of the noble art of applause-catching. He is a capable pianist, but not a virtuoso; here, again, he fails to capture the imagination of the ordinary audience.[37]

Others discovered different qualities in Bartók's playing. Edwin Evans certainly recalled no reticence: 'the energy of some of the piano pieces, played as he played them, does occasionally suggest a swashbuckler at court.'[38] Anthony Clyne went further, praising Bartók's 'superb technical execution', which allowed the essentials of the music to be revealed with subtlety. [39]

From the few who even mentioned it, the opening violin sonata by Mozart elicited compliments. 'Altogether admirable reading', wrote Calvocoressi in the *Glasgow Herald*;[40] 'a delightful performance', commented the critic for the *Liverpool Courier* (the only one to recall attending

---

[36] 'Music of the Week', *Observer*, 26 Mar. 1922.
[37] 'The Week in Music', *MG* 30 Mar. 1922. Newman's view was echoed by Hans W. Heinsheimer many years later: 'Béla Bartók's bows were certainly a concert manager's nightmare: stern, professorial, unsmiling to the extent of chilliness . . .' (in Vilmos Juhász, *Bartók's Years in America* (Washington DC, 1981), 83).
[38] 'Béla Bartók', *Pall Mall Gazette*, 25 Mar. 1922.
[39] 'Béla Bartók and Magyar Music', *Bookman*, 62 (1922), 145–6.
[40] 'The Week's Music in London', *Glasgow Herald*, 29 Mar. 1922.

Bartók's early Manchester performances).[41] For E. A. Baughan of the *Saturday Review*, too, the performance of the sonata was excellent, but he considered that Bartók had an ulterior motive in placing this piece first in such a programme: 'Having made up his mind that clash of sonorities is the soul of music, no matter how ugly it may be, he is not to be seduced from his ideal. At his concert last week he even employed Mozart's Violin Sonata in D as a foil. It was as if he were determined that his pianoforte pieces should seem as ugly as possible.'[42]

With the second offering of the programme, consisting of Bartók's *Suite* and *Improvisations*, the criticism began in earnest. The latter composition, written in 1920, drew most of the comment, as it was considered more indicative of the composer's current attitude. In the *Musical Times* Calvocoressi pointed out with more excitement than accuracy that *Improvisations* was practically the only work in which Bartók went beyond simple, effective settings of folk-tunes.[43] Edwin Evans also found the work significant in terms of the direction of Bartók's output, and placed it second only in interest to his Violin Sonata.[44] For Ernest Newman this significant change in direction showed too many signs of 'dry abstraction' and not enough of imagination. Newman continued: 'At his best he is admirably terse and epigrammatic; his finest music has the spareness and sinewiness of the athlete. But of late the sinewiness seems to be degenerating into stringiness, and the spareness into angularity.'[45] Others expressed their feelings more directly. The *Liverpool Courier*'s critic found Bartók's *Improvisations* 'extraordinary to the point of the Macabre, and after hearing them twice I am none the wiser nor in the least impressed'.[46] In an article entitled 'A Hungarian Genius?', E. A. Baughan confessed that he was perplexed by Bartók's two piano works, and even questioned why such works had ever been composed: 'Clever composers often improvise the strangest music as a kind of protest against the conventions of art, but they do not write their experiments down and certainly do not play them in public. But Bartók takes himself very seriously.'[47]

The four Hungarian folk-songs which followed were perceived as being more conservative in idiom and consequently more acceptable. In them 'a decided character that won the readiest recognition' was discerned by the *Morning Post*'s writer.[48] Even the hostile critic from the *Liverpool Courier* (who left at the interval) managed to find one of the songs 'understandable and pleasant'.[49] In comparing these songs with Bartók's piano works,

[41] Anon., 'London Letter', *Liverpool Courier*, 25 Mar. 1922.
[42] 'A Hungarian Genius?', *Saturday Review*, 1 Apr. 1922, p. 331.
[43] 'Béla Bartók', MT 63 (1922), 344.
[44] 'In London', MNH 1 Apr. 1922, p. 406.
[45] 'This Week's Music', ST 26 Mar. 1922.
[46] *Liverpool Courier*, 25 Mar. 1922. This reviewer had heard the *Improvisations* at the reception on 19 March.   [47] *Saturday Review*, 1 Apr. 1922, pp. 331–2.
[48] Anon., 'Music', MP 25 Mar. 1922.   [49] *Liverpool Courier*, 25 Mar. 1922.

Edwin Evans found gentler harmonies, which helped to create a delightful performance.[50] 'That would not prevent some of our folk-song enthusiasts from crying out aloud', he added, 'if any of our composers treated our folk-tunes with the same freedom.' Lack of appreciation of that freedom led 'Crescendo' from the *Star* to comment:

I heard some Hungarian folk-songs, arranged by Mr. Bartók. His method of arranging accompaniments may be gathered from the fact that if the composer had not been at the piano himself I should have wagered that each song was being sung with an accompaniment meant for another. Still, one of them was encored, principally because Miss Grace Crawford sang it with irresistible charm. Incidentally, being a good patriot, she conscientiously murdered the German language.[51]

With Bartók's Violin Sonata, after the interval, the critics were more occupied. Although traditional in the fast-slow-fast pattern of its movements, the sonata's avoidance of clear repetition of themes caused much soul-searching, especially among these critics who had not attended the private rendition ten days previously. Regarding the actual performance, most opted out of any judgement. 'Crescendo' probably spoke for many: 'Mme. [sic] Jelly d'Arányi struggled heroically with the violin part, and the composer at the piano seemed to be enjoying himself.'[52] Leigh Henry from the *Chesterian* had more firm views, however. He wrote that the sonata had been 'rendered with fine insight and spirit by Jelly d'Arányi, and exquisitely balanced by the composer's piano part'.[53]

The search for a key to elucidate the work's construction occupied most of the critics for paragraphs. Many, failing to find one, took refuge in the paraphrasing of clichés then circulating about Bartók's music. It had to be considered only on its own terms, they advised, although that did not mean one had to like this music. It may well sound complicated, several critics stated, but was essentially simple and direct. In fact, Bartók's very lack of circumlocution was, in their opinion, what so irritated many in Bartók's audience. Only the senior critics were prepared to be more definite than this. Concerning the use of dissonance in the sonata, Eric Blom wrote:

This has been called an ugly work, chiefly on account of the minor seconds and augmented octaves which Bartók uses as freely here as Brahms uses thirds and sixths; but even granted that these intervals are ugly—which is after all a question of usage—the ideas exposed by means of them are surely among the loveliest things music was ever called upon to express.[54]

Whether conditioned by usage or not, Ernest Newman did not like this dissonant work. For his *Sunday Times* readers he penned the memorable

---

[50] *Pall Mall Gazette*, 25 Mar. 1922.
[51] 'A Musical Extremist', *Star*, 25 Mar. 1922.
[52] Ibid.
[53] 'London Letter', *Chesterian*, NS no. 23 (May 1922), 209.
[54] 'Studies at Random', MO 45 (1921–2), 696.

lines: 'The bulk of the sonata seemed to me the last word (for the present) in ugliness and incoherence. It was as if two people were improvising against one another.'[55] Writing again later in the week, Newman still held this opinion, but looked further into the cause of his dissatisfaction.[56] He pursued the theme of excessive abstraction. As with Schoenberg, Newman believed, Bartók was seeking to 'quintessentialise the expression', thereby squeezing the vital element out of music. 'There comes a point', he concluded, 'at which compression can lead only to incoherence, and this is the point, I think, at which Bartók has arrived in the Violin Sonata.'

Leigh Henry and E. A. Baughan paid considerable attention in their reviews to the contrasting roles of violin and piano in the sonata, but for different purposes. Henry congratulated himself for having discerned such a trend in instrumental composition even before the war, and rounded on the academics who had then scoffed at his idea.[57] For Baughan, who had been quite happy to condemn as 'meaningless' the Bartók works presented earlier in the programme, this sonata proved more difficult to dismiss. In it, he believed, something of Bartók's method could be ascertained:

Throughout the sonata I was impressed by the fact that the violin was trying to say something in contradiction to the piano. The sonata is not a fusion of both instruments, but is a sort of duologue: impassioned on the part of the violin; restrained, logical and rhythmical on the part of the piano . . . Puzzling as this sonata was at a first hearing, it was sufficiently arresting in its quality of feeling to make one think that, after all, Béla Bartók may be something of a genius. There is, at any rate, one clear point for praise, and that, strangely enough, is a matter of workmanship. I liked the way in which the composer has kept the character of the violin and piano distinct.[58]

This was not the only persistent contradiction perceived in the piece. A number of the reviews raised the issue of the radically different effects of Bartók's use of rhythms and pitches. While his savage rhythms were found to be impressive, their effect was undermined by the chaotic use of pitches.[59]

Bartók must have been surprised at the caution evident in so many reviews of the sonata. More educated in the techniques of contemporary European composition than had been the case before the war, most critics were conscious of a need for rehearing and reflection before evaluating so significant a work. In the meantime the old temptation of pejorative rhetoric had to be eschewed. For one critic, however, the process of rehearing and reflection had already taken place. Having heard the sonata at the private

[55] ST 26 Mar. 1922.
[56] MG 30 Mar. 1922.
[57] Chesterian, NS no. 23 (May 1922), 209.
[58] Saturday Review, 1 Apr. 1922, pp. 331–2.
[59] See: Daily News, 25 Mar. 1922; MP 25 Mar. 1922.

concert on 14 March, and, one suspects, closely studied the score, Calvocoressi was able to write a truly informed review far surpassing the hurried observations of his fellow critics.[60] He fearlessly tackled the difficult issues of form and tonality avoided by others, finding standard underlying forms to each movement and a consistent approach to tonality. Bartók used neither traditional keys nor polytonality, he asserted, but, rather, polarized each section of his music around a particular note or interval. Calvocoressi's Glaswegian readers (who had not had any opportunity of hearing the work) were probably bemused by the listing of tonal areas which followed. Despite the technical details of his review Calvocoressi made no attempt to hide his admiration for Bartók's progressive work:

The second movement, a most impressive Adagio, consists almost entirely of a long cantilena of the violin, sustained and glowing, which affords a perfect instance of pure melody entirely free from melodic common-places. . . . when a composer succeeds in giving us melody of the finest order whose appeal does not depend in the least upon the familiarity of its various patterns and curves, upon the long-tested reactions of certain progressions and cadences, we ought to feel duly grateful. That Adagio, I believe, ought to be described as one of Bartók's greatest achievements and one, more generally speaking, of unusual greatness.

The short selection of Bartók and Kodály piano pieces with which the concert ended received scant attention in the reviews. Only for 'Marcato' of the *Evening News* was this a highpoint of the concert.[61] At last he had found something which he could understand. Writing a few short paragraphs under the heading 'Mystery Concert', he told of an incident between a scoffer and an enthusiast during the concert, and the gradual departure of less-dedicated members of the audience during the 'extraordinary evening'. 'Marcato' had sympathized with them—until the final bracket of pieces: 'They were wrong—they missed some comprehensible lyrical piano pieces by Kodály and two of Bartók's rhythmical pieces, "Burlesque" and "Rumanian Dance", of a rhythm so taking that we forgot our distress at Bartók's tunes.'

Final judgements of the evening were, on balance, mildly positive. Everything about Bartók's music was heavy going: to play it, to listen to it, and certainly to comprehend it. But it was worth the effort, many critics decided, for it showed too many signs of talent to be ignored or summarily dismissed. The most widespread fault detected was the over-academic attitude to his art adopted by Bartók. E. A. Baughan neatly expressed this belief: 'much of the composer's strangeness of workmanship is due to the cold self-consciousness of a theorist, and not to a tone-poet's instinctive

[60] 'The Week's Music in London', *Glasgow Herald*, 29 Mar. 1922. A summary appeared as 'Béla Bartók', MT 63 (1922), 344. Many of Calvocoressi's ideas were reported by L. Dunton Green in his 'Lettre de Londres', *Revue musicale*, 3. 7 (May 1922), 168.
[61] *Evening News*, 25 Mar. 1922.

search for the medium through which he can best express himself.'[62] For his rhythmic vitality, directness of expression, sheer dynamism, and (with some dissension) beauty of melodic invention, however, Bartók had received due recognition. His cause had prevailed, for the moment, before some of the world's most uncompromising critics.

Reviews aside, other indications of musical success were unequivocally present. After the concert, well-wishers thronged the backstage areas—to history's benefit, as it forced less persistent members of the audience to express their admiration in writing. The Wilsons, who had provided accommodation for Bartók earlier in the week, congratulated him on the great success of the evening: 'You were such a LION last night', wrote Freda Wilson, 'that there was not room for Duncan at all in the green-room!'[63] The young British composer, Arthur Bliss, then at the height of his interest in the latest trends of European composition, felt compelled to pen his views on the sonata, since he had not been able to speak with Bartók in person. He found it a remarkable work, which would certainly gain Bartók many followers.[64] For Bartók, however, the most exciting result of the evening, and surest sign of recognition, was the emergence of plans for a return visit to London in November.[65] Only by repeated exposure, he knew from past experience, would his music move from novelty to mainstream status.

Although busy with his own concert engagements, Bartók was intent on hearing all he could of the work of other composers and performers. On 20 March he heard the London Symphony Orchestra at the Queen's Hall and departed with mixed impressions. While full of praise for its string section and the conductor Albert Coates (who had been Director of the Imperial Opera in St Petersburg before the Revolution), he considered the wind players inferior to their counterparts in Budapest.[66] In the programme he was most taken with the final movement from Stravinsky's *Firebird* suite (1911), because of its rhythmic drive and 'Eastern freshness', but considered Respighi's *Ballad of the Gnomes* (1918–20) 'rather weak'. Three days later he attended the Royal Philharmonic Society's première of Delius's *Requiem* (1914–16), but found it less appealing than that composer's earlier works— in a word, 'shallow'. The little other contemporary English music which he heard at this time left him similarly unimpressed. Bartók's musical horizons were broadened in other ways as well. Through a meeting with the harpsichordist Violet Gordon Woodhouse he came to know more about the capacities and playing techniques of that instrument. He was surprised to hear some of his own piano pieces for children played in this different medium. Curiosity also drew him to the Pianola Company's music-room,

[62] *Saturday Review*, 1 Apr. 1922, p. 331.
[63] Unpublished letter, 25 Mar. 1922, in English, MTA BA-B 3477/234.
[64] *DB*.iii pp. 114-15.   [65] *BBlev*.v p. 280.
[66] Aladár Tóth, 'Bartók külföldi útja', *Nyugat*, 15 (1922), 830–3 (*ZT*.vii pp. 219–22). This article is the source for the remainder of this paragraph.

where he discussed with Edwin Evans the potential of that instrument for creating an 'objective music'. While expressing interest in Stravinsky's use of the pianola in recent works, Bartók still harboured doubts about the instrument's relevance to serious composition.

Keen to meet the man who had done so much to arrange British premières of his orchestral works, Bartók visited the Queen's Hall on Sunday 26 March and showed Sir Henry Wood the score of his Four Orchestral Pieces, Op. 12 (1912, orchestrated 1921). These pieces had received their first performance in Budapest only two months before. In a subsequent letter to Bartók, Wood stated his intention of conducting the work during the next season, although in the end no performance took place.[67]

Not all Bartók's spare time was taken up with musical activities. He was anxious to enjoy those luxuries which were part of the everyday life of his London hosts, but quite beyond his means in Hungary. Writing to his 11-year-old son Béla on the reverse side of a postcard of Westminster Abbey, he told of two gastronomic forays:

26 March 1922. London.

My concert was the day before yesterday. Afterwards someone took me to have supper—just imagine with whom—the famous Marconi, who was throwing a big party in a hotel. (It was only completely by chance that I dropped in there.) There were all kinds of good things there: oysters, fish, game stuffed with goose-liver, champagne, real cognac. But you would have stayed hungry! Still, the day before yesterday I was taken to a Chinese restaurant. Of course I wanted to order dog and cat meat (if it is going to be Chinese, then let it really be Chinese), but there was nothing like that on the menu. But I still ate some rather curious things. Horses have already completely disappeared from the streets here; just once in a while you see one or two harnessed to a carriage. Bye-bye—and kisses,

your Father[68]

After some days staying in central London just off Piccadilly, Bartók travelled to Liverpool on Thursday 30 March for a concert before the British Music Society in Rushworth Hall. True to his offer of support of the previous December, Eaglefield Hull had initiated the idea, although not without an element of self-interest. As editor of the forthcoming *Dictionary of Modern Music and Musicians*, he needed a willing contributor from Hungary, and needed him soon, as most other national authors had already been arranged. Having provided Bartók with an additional venue for performance, Hull moved quickly, on 28 March offering him the responsibility for all articles on Hungarian subjects, at the modest rate of 2 guineas per thousand words.[69] Equally speedily Bartók accepted—2 guineas

[67] DB.iii p. 115.
[68] BBcl p. 239. Guglielmo Marconi was the inventor of wireless.
[69] MTA BA-B 3477/91.

would mean a lot back in Hungary—and drew up a list of possible entries for the dictionary on the back of his Liverpool programme.[70]

In Liverpool, Bartók performed twelve of his own pianoforte compositions, grouped within three brackets. Individual items varied considerably in technique and length, from the four-movement *Suite*, Op. 14 to the short pieces of 'Evening in the Country' (1908) and 'Bear Dance' (1908).[71] Eaglefield Hull introduced the programme with a short description of Bartók's use of folk-elements in his compositions, but reportedly did little to enlighten his audience. By and large, the criticisms of this Liverpool concert were the most candid 'first impressions' from Bartók's 1922 tour. They lacked the rustic amazement exhibited by the Aberystwyth amateur commentators, but also avoided the self-consciousness and sensationalism of many of the London professionals. These Midland critics were prepared to acknowledge their own ignorance, but still tried hard to analyse the music before coming to their frank, mostly unenthusiastic, conclusions.

In *Musical Opinion* the Liverpool correspondent wrote of the pride of the local branch of the British Music Society in securing Bartók's services, which had turned to dismay on hearing his works: 'Well, M. Bartók came, astonished where he did not baffle, and departed without leaving behind too much conviction.'[72] More personally, this critic considered Bartók's harmonies too often 'arbitrary and harsh'. Thereby his enjoyment of most of the evening's works was destroyed. These views were echoed more thoughtfully and mildly by A. K. Holland of the *Liverpool Daily Post*, who expressed an inability to follow the longer, more complex pieces. He found the most likeable performances in the simple 'Bear Dance' and in a short encore selection of pieces from *For Children*. But Holland was never dismissive, and took pains to describe Bartók's technique of composition:

His piano music is spare in substance, stark in outline. There are whole stretches of pure line in octaves. One noted a certain lack of harmonic colour, a preference for the bare chord. Two-part contrapuntal passages abounded, and there was a wiriness in the contour of the music which left the shape clear and emphatic, even when it seemed to lack significance of form.

The *Evening Express*'s critic recorded the perfunctory nature of the applause, which, he suggested, resulted partly from the tedium of any 'one-man show', but also came about because the audience found the music unintelligible and bizarre. While some might like their music in Bartók's fashion, this writer sighed, he preferred the virtues of old-fashioned Hungarian music: sparkling melody, capricious tempos, and passionate

[70] MTA BH 2400/497.　　　　　　　　[71] MTA BH 2049/95 and 2400/497.
[72] Reviews mentioned: Anon., 'Music in Liverpool', MO 45 (1921–2), 685; A. K. H., 'M. Béla Bartók', *Liverpool Daily Post*, 31 Mar. 1922; W. J. B., 'Béla Bartók', *Evening Express* (Liverpool), 31 Mar. 1922; T. J. B., 'Hungarian Composer', *Liverpool Echo*, 31 Mar. 1922.

expression. As a pianist, too, Bartók was found wanting. He demonstrated a comprehensive technique, this critic claimed, but lacked the ability to grade his tone.

The Liverpudlians gave Bartók his coolest reception of the tour, but without a hint of venom. The *Liverpool Echo*'s column concluded: 'On the showing of last night one cannot but think that M. Bartók's reputation has been rather exaggerated, but the conclusion is unfair as the programme did not fully represent him.' To A. K. Holland, Bartók's personality was a major impediment to his music's successful propagation:

The composer had slipped away before the applause had ceased, and could not be prevailed upon to continue. M. Bartók, an extremely modest and retiring personality, is perhaps the last person to popularise his own music, which contains considerable vigour and energy, as well as an occasional wit and humour.

Bartók's impressions of the concert are not recorded, but curiously, among his memorabilia is found a printed 'compliments card' of Eaglefield Hull bearing the simple message: 'Bartók was very successful at Liverpool.'[73] Whatever the reasons for Hull's expression of opinion, the critics were certainly united in their view: Bartók had gained few adherents that night.

Returning to London, Bartók took part in a small private concert at the South Kensington home of Duncan and Freda Wilson on 31 March. The programme placed a selection of Bartók's shorter piano works in the setting of Bach's E major and Beethoven's 'Kreutzer' Sonatas, both played with Jelly Arányi.[74] Because of the strictly private nature of the concert there were probably no reviews. The only impressions of the evening are found in a letter of thanks from Duncan Wilson to Bartók:

You probably cannot realise what intense pleasure it was for us and our guests to hear two such artists in our house . . . The Beethoven especially was a revelation to me. I have never heard a performance to approach it, and I only hope I shall have another opportunity of hearing you play it. We both enjoyed having you with us, though we saw too little of you, and we shall not forget that you have *promised* to come again when you [are] here in the autumn.[75]

After a final weekend in central London, Bartók left for Paris on 3 April, accompanied by Jelly Arányi and her mother. Press attention to their concerts together in Paris was less intense than in Britain, although still cautiously favourable. The most important engagement occurred on 8 April, when they performed the Violin Sonata twice at the one reception.[76] Looking back in May on his British and French visits, Bartók considered that they had gone 'like clockwork', in contrast to the shambles surrounding the performances of his stage works in Germany during May.[77] Financially,

---

[73] MTA BA-B 3477/96.   [74] MTA BH 2049/97. See Plate 4.
[75] Unpublished letter, Duncan Wilson to Bartók, 2 Apr. 1922, in English, MTA BA-B 3477/233.   [76] *BBcl* pp. 330–2.   [77] *BBlev*.v p. 281.

too, Bartók could not complain about his stay in Britain: he had made a profit of £82.[78] That surplus would prove invaluable at home, where, during Bartók's absence, his wife had rented a self-contained flat. The value of stable foreign currency was graphically illustrated in August 1922, at which time one week's rental of a respectable apartment in London equalled, at current exchange rates, Bartók's rental for an entire year in Budapest, namely, £2.[79]

Although Bartók's concerts in Britain were over, his profile in the press remained high for some weeks. Hungarian music and musicians were popular topics in many quarters, particularly in the April issues of the journals. Both Edward J. Dent and Leigh Henry wrote general appraisals of Bartók, while Calvocoressi widened the consideration of Hungarian music with an article about Bartók's compatriot Zoltán Kodály.[80] The comparisons which he drew between the two were not always flattering to Bartók. In his use of pattern repetition in working out forms, Kodály was judged without equal. His Solo Cello Sonata, Op. 8 was cited as an example. On the other hand, in the works of Bartók which abandoned many of the conventions of tonal music, Calvocoressi detected an awkwardness in the handling of form. Cryptically, Calvocoressi concluded: 'Hence the necessity, certain courses of procedure being eliminated, to resort to new ones.' A third Hungarian, Ernő Dohnányi, also featured in the British press at this time, because of the gathering in Budapest of many musicians for the première on 18 March of his opera *A vajda tornya* (1915–22).[81]

The journal coverage abated little in May. Cecil Gray's article, 'Zoltán Kodály', again introduced frequent comparisons between the two Hungarians.[82] In his preamble he pointed out that the constant association of the two by music commentators had done much to belittle Kodály's originality and talent. The resemblances were only superficial, Gray asserted. Kodály's work was more uncompromisingly national than Bartók's and, if anything, showed a French orientation against Bartók's more Beethovenian stance. To Gray, 'Bartók touches greater heights and wider extremes than Kodály; his genius is of a more robust order. Kodály seems rather to avoid extremes . . . simply out of a natural and instinctive reticence and a habit of emotional reserve.' Elsewhere, Eric Blom cursorily surveyed Bartók's string quartets, which at that time had still not been heard in Britain.[83] The article was probably a response to advance notices of a

---

[78] *BBcl* p. 329.   [79] *BBcl* pp. 335–6.

[80] Edward J. Dent, 'Béla Bartók', *The Nation and the Athenaeum*, 1 Apr. 1922, pp. 30–2; Leigh Henry, 'Béla Bartók', *Chesterian*, NS no. 22 (Apr. 1922), 161–7; M.-D. Calvocoressi, 'Zoltán Kodály', *MMR* 52 (1922), 79–81.

[81] See *The Times*, 28 Mar. 1922.   [82] *MT* 63 (1922), 312–15.

[83] 'Studies at Random', *MO* 45 (1921–2), 696–7, revised as 'Béla Bartók as Quartet Writer' in his *Stepchildren of Music* (London, 1925).

performance of the Second String Quartet, Op. 17 by the Hungarian String Quartet at the Grafton Galleries on 9 May. Scheduled so soon after Bartók's own visit, this performance drew a good audience. The Second Quartet was, indeed, the work which Cecil Gray had claimed as the greatest quartet since Beethoven.[84] Edwin Evans and Calvocoressi both wrote substantial reviews of the performance. Evans heard a fine rendition of 'healthy, rich music of neo-romantic tendency', with 'every point being made to stand out with the requisite character'.[85] Calvocoressi's *Glasgow Herald* readers received a long description of the work's three movements and a confession that their correspondent did not yet understand too much of the composition's inner meaning.[86] Those who hoped that Bartók's chamber music would provide a simple key to the comprehension of his art undoubtedly left the concert frustrated.

As the months passed Bartók's exposure in the press gradually lessened. Looking back in June on Bartók's visit, Anthony Clyne noted two important results: a reduction in the level of ignorance about Hungarian music, and the arousal of a genuine sympathy in British audiences for some of the works presented.[87] 'The Magyar temperament', he stated, 'is not antipathetic to the British, and to us Magyar music will appeal, when it is understood, more perhaps than to the French, for example. It has nothing of that delicate elusiveness which often becomes just dreamy confusion. It is strange, but not "insaissable".' Reports of the production of Bartók's opera and ballet in Frankfurt-am-Main on 13 May also kept his name before the British public during the summer months, although these reports did not disguise the limited success of that evening.[88] Only in August did the final article inspired by Bartók's spring tour appear, not surprisingly, from the pen of Calvocoressi.[89] Writing 'More about Bartók and Kodály' for the *Monthly Musical Record*, he stressed how unfortunate it was that during the visits of both Bartók and the Hungarian Quartet financial considerations had so limited their public exposure:

... obvious financial reasons compel many of the artists who are visiting London this year (and not a few of our own as well) to give their concerts at private houses instead of renting halls, and to cut down advertising expenses to a bare minimum, if not to forgo them altogether—with the result that even when they play new works, these artists, from the point of view of propaganda and that of education, achieve far lesser results than they would achieve under normal circumstances.

That the works of Bartók and Kodály were worthy of the highest level of public explosure, Calvocoressi had no doubt. In the quality of their melodic

---

[84] 'Béla Bartók', *Sackbut*, 1 (1920–1), 303.
[85] 'London Concerts', *MT* 63 (1922), 412.
[86] 'The Week's Music in London', *Glasgow Herald*, 17 May 1922. See also *Revue musicale*, 3. 10 (Aug. 1922), 169–72. [87] *Bookman*, 62 (1922), 145–6.
[88] See: *The Nation and the Athenaeum*, 3 June 1922, pp. 354–6; *MT* 63 (1922), 490, 513.
[89] *MMR* 52 (1922), 181–3.

invention and their capacity to mould successful forms he now heralded both equally as masters. Because of their steadfastness of purpose, high level of technique, and genuine originality, bright futures were assured for both.

After returning home on 19 May, Bartók was interviewed several times about his ten-week excursion into western Europe.[90] On each occasion he spoke enthusiastically about his journey and the reception of his music. One report by Lajos Kőmives in a Transylvanian newspaper was particularly detailed.[91] Entitled 'My Visit to Béla Bartók', the article began with a finely drawn portrait of Bartók in his forty-second year:

The door quietly swings open and the Master appears in its opening. At first sight he appears unpretentious and unassuming. The theatrical manner of internationally-fêted celebrities is completely absent from his demeanour. He is not robust, is rather shorter than average in stature, and his black clothes vividly set off the outline of his refined white countenance, from which his shrewd, stern eyes, radiating intelligence and ardent interest, attest to the presence of a superior spirit. His face is still thoroughly youthful, but above the smooth high forehead his short hair has already turned completely white. That is not a sign of old age but only of premature greying. It bestows on his whole head a quality of greater whiteness and interest.

In the interview which followed Bartók described his surprise at being so well received by musicians in Britain, and told of the attention which had been lavished upon him in the press. As a Hungarian, too, he felt that he had been generously received, and stressed to readers how the British critics had placed much store on his Hungarianness, referring on a number of occasions to the natural talents of the Hungarian people. Looking back on his weeks in Britain, Bartók retained only the happiest of memories. An uncharacteristic hint of overstatement crept into his final reported impression of the visit:

My public concert in London took place in the biggest and most distinguished of its concert halls. The ovation which bombarded the stage after the completion of my violin sonata was rather unexpected and embarrassing, coming from an English audience normally characterized by its coolness and restraint. From London I went to Liverpool and from there [sic] to Aberystwyth in the principality of Wales. The audience there granted me an ecstatic reception similar to that in London. Wherever I went I only heard favourable things said about the Hungarian people, and you can imagine how well this pleased me out there in that foggy, strange land: 'We don't harbour resentment against the Hungarians. During the war we considered they had been forced to be our opponents and were victims of the Germans.' I heard statements like this, or something similar, countless times on my trip, and every time it seemed that these statements were expressed frankly and not just as an obligation of courtesy.

[90] ZT.vii pp. 218–22.
[91] 'Látogatásom Bartók Bélánál', Keleti Újság (Kolozsvár/Cluj), 31 Oct. 1922 (ZT.vii pp. 234–7).

4.

1923: TOWN AND COUNTRY

In 1922 the novelty of Bartók's music led to indecision among the British
critics; in 1923 familiarity bred scepticism and even hostility. Whereas at
the beginning of the earlier year he was certainly underexposed, by the
conclusion of his second visit of 1923 many British music-lovers were
satiated with his music. The critics turned to pursue other novelties and
sensations, and forsook Bartók's narrow and repetitive concert programmes.
Even at this early stage of international recognition he was regularly linked
with Stravinsky and Schoenberg in a vanguard triumvirate of modern
music, but often this link was used in condemnation, not praise, of his
achievement. The problem for many listeners was simple: they were unable
to dismiss the music of these three composers merely on the grounds of
stylistic dislike, and increasingly found themselves concluding that these
radical compositions were not music at all.[1] The few years of musical
liberality immediately after the war were ended by a conservative backlash
in which Stravinsky's *Rite of Spring*, Schoenberg's atonal works, and
Bartók's violin sonatas were front-line targets for ridicule. The hold of the
conservatives was reinforced by the failure of several progressive music
journals, by falling concert attendances, and through conservative successes
in the fierce debates about the presentation of music on radio. While in
1923 the British became comfortable with the notion of Bartók as a leading
progressive composer, in return they terminated his 'honeymoon' with
many biting criticisms of the very purpose of his music.

Not merely in the consolidation of his international status, but also in
compositional technique and personal matters, 1923 proved an important
year in Bartók's life. His *Dance Suite* for orchestra, written in celebration of
the fiftieth anniversary of the union of Buda and Pest, was his only work for
the year, but a most significant one, for it introduced a more simple and
popular style of writing which drew away from the harsh dissonances and
atonal tendencies of the violin sonatas. Despite a poor première the work
soon became Bartók's most performed composition. During the 1925–6
season it was played in over sixty cities around the world, including multiple
performances in London and Birmingham. Emotionally, too, 1923 was a
momentous year for the composer. He divorced his wife Márta, with whom
relations had been strained for some time, and immediately married a young
student, Ditta Pásztory.[2] The close personal and musical collaboration with

---

[1] See, e.g. R. C. [Capell], 'Music Notes: Is it Music?', *DM* 10 Dec. 1923.
[2] See István Volly, 'Bartókné Pásztory Ditta', *Életünk*, 21 (1984) 809–10.

violinist Jelly Arányi also came to an end, although the two did perform together in isolated concerts in later years.[3] Perhaps because of his personal turmoil Bartók proved even less communicative than normal about his private thoughts when corresponding with family members. Several passages from his few letters from Britain during this year were little more than financial statements.[4]

1923 was also the year in which Bartók became involved with Oxford University Press. On 7 May terms were agreed between Bartók and OUP for the publication of the volume *Hungarian Folk Music*.[5] Later, in 1925 and 1926, arrangements were made for his collection of Romanian Christmas Songs to be published, and it was suggested towards the end of the decade that OUP might also become the publisher of his compositions. Because of the General Strike of 1926, difficulties with translators, and a cooling of relations between Bartók and Hubert J. Foss of OUP's Music Department, the Hungarian volume only appeared in 1931.[6] Plans for the Romanian book finally had to be abandoned in 1935.

Personal impressions of Bartók, the man, as distinct from his music, were rarely recorded by British music commentators during 1923. One critic, Watson Lyle, who had not attended any of Bartók's concerts in 1922, did however provide a fine verbal portrait.[7] Extrapolating from Bartók's music, Lyle had expected to meet a cynic, and so he was surprised by Bartók's personal sincerity and tenderness:

There is in his nature the charitableness, the warm humanity, and the clarity of thought that are antagonistic to the harbouring of jaundiced views upon either Life, or its reflection, Art. . . . His humoursome smile, and simple directness in expressing an opinion, are alike free from pose. Slight, physically; unassertive yet dignified in manner, his dark eyes often glow with a compelling intensity in his sharp-featured face. His hair is white. One feels, however, that he is of the eternally young in spirit . . . It hardly seemed possible that the composer who strongly, sometimes almost defiantly, proclaimed the distinctive individuality of his art, and himself, when seated at the keyboard, could be the modest man I met the previous day who, when I referred at all to his music, was much more inclined to change the subject than to discuss it.

Bartók's two visits, in May and November–December, were the most varied of his many British tours. Their study reveals much about British music-making on the verge of the era of radio. Although based in London, he journeyed from Malvern to Aldeburgh, Battle to Huddersfield, in a range of public, society, and private recitals arranged by his British supporters. He played in solo, chamber, and orchestral soloist roles, always featuring his own compositions. As in 1922, these concerts were organized only slowly,

---

[3] For details of their concerts in 1928 and 1930 see Chapter 8, below.
[4] *BBcl* pp. 340, 345.   [5] MTA BA-B 3477/82.
[6] See Malcolm Gillies and Adrienne Gombocz, 'The "Colinda" Fiasco: Bartók and Oxford University Press', *ML* (forthcoming).   [7] 'Béla Bartók', *MNH* 19 May 1923, pp. 495–6.

causing Bartók to delay the final decision about his trips until all costs were amply covered. The original plan, formed at the height of his London success in March 1922, had been for him to revisit Britain in November of that year,[8] but by June it had become clear to him that the Arányis were doing little to secure engagements and, as the year progressed, that his agent at E. L. Robinson Concert Direction could produce no definite plans.[9] It looked for several months as if 1922 would prove to be yet another false start to enduring British recognition. But then, in March 1923, at the suggestion of its president, Edward J. Dent, the British Office of the newly formed International Society for Contemporary Music wrote inviting Bartók to perform a programme of his own works at its headquarters in London.[10] Two other engagements were arranged at this time, all conveniently falling in early May immediately following a tour of Holland to which Bartók was committed. While he was actually in Britain two more recitals were offered. Of these five performances, presented between 4 and 14 May, two were private engagements, and three were before associations which allowed access to the public.[11] With the November–December tour Bartók was similarly cautious, agreeing to travel only when four concerts had been fixed, but eventually playing at eight or nine following last-minute, often lucrative invitations to perform privately.[12] These would be the last formal private concerts at which Bartók would perform in Britain. When he returned four years later new advances in public entertainment would have made private concerts less attractive both to audiences and artists.

'In England you can give recitals at beauty spots like this village', wrote Bartók to his mother on a picture postcard from Malvern Wells. 'I am playing here today—at a girls' school!!'[13] In such a surprising place Bartók began his first tour of 1923. After two similar school appearances he would

---

[8] BBlev.v p. 280.   [9] BBrCal pp. 212–17.

[10] MTA BA-B 3477/207. The Society came into being in January 1923 after a congress in London.

[11] 4 May: The Abbey School, Malvern Wells (private), solo recital; 7 May: International Society for Contemporary Music, London (public access), sonata evening with Jelly Arányi; 9 May: British Music Society, Huddersfield (public access), solo recital; 11 May: George Woodhouse Pianoforte School, London (public access), solo recital; 14 May: Normanhurst Court (a school), Battle (private), solo recital.

[12] 30 November: Aeolian Hall, London (public), recital with Jelly Arányi; 1 December: probably a private concert in London, with Adila Fachiri; 3 December: Fachiri home, London (some public access), recital with Adila and Alexandre Fachiri; 4 December: Belstead Hall (a school), Aldeburgh (public access), solo recital; 6 December: Winter Gardens, Bournemouth (public), with Bournemouth Symphony Orchestra conducted by Sir Dan Godfrey; 7–9 December: at least one private concert, London, possibly at the Mayer/Moulton home, with Jelly Arányi; 10 December, George Woodhouse Pianoforte School, London (public access), recital with Adila Fachiri; 12 December: St James's (a school), West Malvern (private), solo recital.

[13] Postcard, Bartók to his mother, 4 May 1923, in Hungarian, reproduced in English in BBL p. 162.

finish his second tour in December at a sister school nearby. These school concerts were arranged by George Woodhouse, a leading London piano-teacher who was involved, along with his wife, in music-teaching at all four schools. In London, too, Woodhouse arranged two concerts for Bartók at his Wigmore Street studios. His interest in Bartók's music initially lay in the easier pre-war piano pieces, which he found ideal for his teaching purposes. Having met Bartók in 1922 and heard him play, Woodhouse became more broadly captivated by his music. At the student concerts of his piano school, works such as the *Elegies*, Op. 8*b*, *Burlesques*, Op. 8*c*, Sonatina (1915), and 'Bear Dance' (1908) became common items.[14] Woodhouse saw that by helping Bartók with concert venues he would further both their interests.

On 4 May Bartók performed at The Abbey in Malvern Wells, probably in the early afternoon. As was often his custom, Bartók 'warmed up' with three short Scarlatti pieces before introducing a series of his own piano works, including the last nine of his *Fifteen Hungarian Peasant Songs* (1914–18) and some favourites—'Bear Dance', *Allegro barbaro*, 'Evening in the Country'—probably known to more advanced pianists among the students. In the second half he performed six Debussy pieces, mostly *Préludes*, followed by further compositions of his own, such as the *Romanian Folk Dances* (1915) and Sonatina.[15] The responses of the students and staff have not been recorded, although George Woodhouse did write at length about the visit for the summer issue of the school's magazine the *Hazel Nut*.[16] Perhaps stung by a sluggish response, he stressed the importance of the occasion, the honour to the school, and the great future which many predicted for Bartók. He then gave a brief description of Bartók's art, including this beautifully sharp passage:

He [Bartók] has discovered new ground, remote from our crowded towns and modern civilization. His music voices a new exhilaration of the human spirit. It breathes the atmosphere of higher altitudes, and supplies the right anti-toxin to those minds which are a little jaded by the familiarity of nineteenth century romantic music. His harmonies and rhythm have none of that cloying character which soon palls on the senses; they are keen and invigorating to all who can respond to 'plein air' music. It is not for musical invalids, for such as desire nothing better than that product of debased art, the sentimental ballad, and its counterpart, the 'drawing room piece'. His rhythms are sharp-edged; Bartók touches a demisemiquaver as with the point of a rapier; and the piano, with its genius for sensing the very pulse of music, is his instrument. . . . Children of today have no taste for music intended as a nursery pastime, and tire soon of the manufactured 'fairy tale' melodies, but here is real music to be remembered a lifetime, and always with a new interest.

[14] Programmes of several of Woodhouse's student concerts during 1922–5 were among Bartók's personal papers (MTA BH).
[15] MTA BH 2049/115. See Plate 5.
[16] 'Béla Bartók', *Hazel Nut*, Summer 1923. Most of this essay was reproduced in *MNH*, 1 Dec. 1923, p. 474.

After the concert Bartók stayed overnight in Malvern Wells at the May Place Residential Hotel, taking tea and supper there, probably alone. He was not given celebrity treatment. Indeed Bartók, whose motto was 'exactitude above all', did not even quibble with his registration under the name of 'Mr. Barkup'.[17] Only two comments emerged from Bartók's pen about his stay in Worcestershire. Writing several days later to his (first) wife from the 'horrible industrial town' of Huddersfield, he mentioned that Malvern was certainly a more beautiful place, but he was more concerned with a 'pleasant surprise' there: a fee of 25 guineas instead of the expected 15.[18] Perhaps influenced by this windfall, the usually cautious Bartók indulged in a full formal luncheon in the 'Hereford Car' during his return train trip to London on the following day.[19]

The hope of recognition rather than reward drew Bartók within a few days to his second provincial engagement, in 'horrible' Huddersfield. A. Eaglefield Hull, who lived in the town, had arranged for Bartók to perform on the evening of 9 May before the local chapter of the British Music Society at Highfield Hall. The fee was 10 guineas for an identical programme to that of 4 May.[20] According to the unnamed critic of the *Huddersfield Daily Examiner*, the concert was very well attended and, in contrast to a similar solo recital of the previous year in Liverpool, it was warmly received.[21] About the events of the evening this critic wrote a substantial column full of fresh and perceptive opinions. He began by asserting that Bartók's music was most individual and most vital, but consequently very easy to misunderstand. His particular worry about comprehension came not from Bartók's dissonances, which he considered par for any course of modern music, but from his nebulous forms:

Now and then one had an uneasy feeling that the whole music was being improvised there and then. The form was frequently so loose that it was difficult to bring the whole into perspective. There is, of course, nothing sacred or fixed in form. It is not essential that good music should follow fixed rules of development, but it is essential that music should possess ideas, that the ideas should progress, and that their progression should be clear.

To this critic the Debussy *Préludes* which Bartók performed revealed a clear progression of ideas, whereas many of Bartók's own pieces appeared, on a first hearing, merely to wander. As a composer, however, Bartók still merited praise for his command of pianoforte colour, rhythmic and melodic invention, and for his ever-present 'cleverness'. As a pianist he was deemed

[17] Hotel bill, MTA BH D-III.12 Számlák.
[18] *BBcl* p. 340.
[19] Luncheon menu, 5 May 1923, MTA BH D-III.13 Számlák.
[20] *BBcl* p. 340, MTA BH 2049/117 respectively.
[21] Anon., 'Mr. Béla Bartók', *Huddersfield Daily Examiner*, 10 May 1923.

'incisive and sensitive, brilliant rather than subtle, and crisp rather than fluent'.

Again Bartók returned to London for a recital before venturing out for his third concert in the country. This had been arranged at short notice. Bartók had expected to leave Britain on 13 May, but on 9 May reported home: '. . . on the afternoon of Monday 14th I am going to play at some school somewhere in Sussex, but I don't yet know for how much'.[22] A subsequent letter on 12 May still concealed the exact venue, but not the fee, of 15 guineas.[23] On the day, Bartók took the train to Battle in East Sussex for an afternoon concert at Normanhurst Court, a small school for girls.

The school was in many respects typical of those visited by Bartók in 1923. Its enrolment at this time was about one hundred, consisting, in the words of one former pupil, of 'daughters of the very titled (or the very rich sent to pick up the King's English from the former), with a makeweight of poor relations at cut prices to fill the pews in church on Sundays'.[24] Originally founded in nearby Bexhill-on-Sea at the end of the war by Margaret Battine, the wife of an army major, the school had soon been moved inland to Battle, to occupy a large rented mansion. This house had been built in 1867 for the Brassey family, of railway-contracting fame, and featured turrets and towers (one of six storeys), rising above terraces flanked by wistaria-covered balustrades. Inside there was a huge front hall (with a branched staircase leading to broad galleries), magnificent reception rooms, and a labyrinth of smaller rooms. Outside among the flower-beds were found various water features: fountains and fishponds. Musically the school was not renowned, lacking an orchestra, choir, or famous ex-pupils. One student who attended the school in the mid-to-late 1920s recalled:

There were so-called 'Musical Evenings' when some of us played our pieces or sang our songs in the drawing room while others rolled lace onto satin camisoles. Mrs Battine, totally unmusical, clapped ecstatically, and the Major always walked out in disgust. However there were occasional recitals. I remember Ruth Gourlay, pianist, and the Budapest Quartet (because of Bartók), and, occasionally, George Woodhouse, who also gave some of us annual 'Master classes on the piano'.[25]

To this institution came Bartók in search of his fee. The programme, Roneoed in blue ink because of the last-minute engagement, was the same as that heard in Malvern Wells and Huddersfield, except for the omission of two short items.[26] Details of the concert's reception are lost, but Iris Gibb, a student who later became a piano-teacher at the school, recalled sixty years

---

[22] Letter, Bartók to Márta Ziegler, 9 May 1923, *BBcl* p. 340.
[23] *BBcl* p. 340.
[24] Unpublished notes to the author provided by Verily Anderson (Cromer, Norfolk), 12 July 1983. See Plates 6 and 7.
[25] Ibid.   [26] MTA BH 2049/119.

later: 'Yes, I was at the old St. Hilary's among the "makeweight of poor relations" to fill the Church pews. Yes, I remember Bartók coming to play, a strange little man with enormous hands, playing (to us) utterly incomprehensible music.'[27] One can wonder what Mrs Battine made of it, or the Major.

After the recital Bartók returned immediately to London and left on an evening train for the Continent. In just under a fortnight he had received £83, of which £65 remained as profit. Most of this came from the concerts arranged by George Woodhouse.[28] The Woodhouses had even organized a dinner in Bartók's honour, which was held in London at the Red Lion Restaurant on 12 May.[29] On the reverse side of Bartók's menu the guests signed their names. Apart from the evening's hosts they included the pianist Katherine Goodson Hinton and her husband, the composer Arthur Hinton; the conductor Anthony Bernard; and the German musicologist Alfred Einstein. Naturally, when Bartók returned to Britain later in the year he was keen to re-establish his links with George Woodhouse. This renewed collaboration resulted in the two further girls' school performances and an additional studio concert in London.

The first of these later school recitals took place on 4 December at Aldeburgh, on the coast of Suffolk. At the time the town had no significant reputation as a musical centre. Within Belstead Hall, however, around eighty girls received a progressive education including systematic music training. The school, for young ladies of 'breeding and means', had been founded in 1906 by Mrs Griselda Hervey, a member of a well-known Suffolk family. One 'old girl', who entered the lowest form in 1923, recalled:

Mrs Hervey had very strong views on what was important in female education. We had wonderful concerts, lectures and visitors; and Cecil Sharp came down every Summer Term with members of the English Folk Dance Society to teach us his 'new' collected country songs and dances. Folk songs of his were sung for fifteen minutes by the whole school every day after prayers. . . . All the top music pupils (piano) were taught by Mr G. Woodhouse who came from London one day a week, and who was much revered . . . Music was very important and a very good school orchestra was in being.[30]

To this enlightened provincial audience Bartók presented the programme reproduced below.[31] Members of the public were invited to attend, although no local music critics appear to have availed themselves of the opportunity.

[27] Unpublished letter, Iris Gibb (Henfield, Sussex) to Verily Anderson, 9 July 1983.
[28] See *BBcl* p. 340.
[29] Menu, MTA BH D-III.95 Számlák.
[30] Unpublished letter, Margot Hare (Saxmundham, Suffolk) to the author, 6 July 1983.
[31] MTA BH 2049/131.

BELSTEAD HALL, ALDEBURGH

PIANOFORTE RECITAL

By

BÉLA BARTÓK,

TUESDAY, DECEMBER 4TH,
at 3 o'clock

Tickets 5/- Reserved Seats. To be obtained from Mrs. Dudley Hervey, Westfields.
"   2/6 Unreserved. To be obtained at the door.

PROGRAMME

I. Scarlatti. (a) Three Piano Pieces.
   Bartók. (b) Tema con Variazioni.
          (c) Old Dance Tunes.
             (from '15 Hungarian Peasant Songs')

II. Bartók. (a) Bear Dance.
         (b) Allegro Barbaro.
         (c) Evening in the Country.
         (d) Three Burlesques.
            (A Quarrel, Un Peu [sic], gris Molto ed priccioso [sic].)

III. Debussy.   'Pour le Piano'
              Prélude—Sarabande—Toccata.

IV. Bartók. (a) Roumanian Peasant Dances.
         (b) Dirge.
         (c) Sonatina.
         (d) Roumanian Dance.

How did Bartók appear to those young ladies of Aldeburgh? Margot Hare, then aged 11, recalls: 'Bartók's concert was in the afternoon—I think on a Saturday. The whole school and staff attended, also some outside visitors. Bartók seemed a very vigorous player, and several of us remember that he blew his cheeks in and out while playing.'[32] And how did Bartók find his young audience? Fortunately for us, on that day he was feeling guilty for not having written to his mother since arriving in Britain, and so

[32] Unpublished letter, Margot Hare to the author, 6 July 1983. The concert was actually on a Tuesday.

before the concert started he whiled away the time by commencing a letter to her:

It is always so worthwhile coming to England. When I set out I was only going to perform 4 times, but it turns out now that I shall play 8 times. I'm already over two of them. Today is the third. (Actually I'm not writing this letter in London, but in a little seaside town, where I shall play at a girls' school at 3 p.m.) While waiting for the concert to start I'm writing letters. . . .———(this long line means that the concert has now taken place. 100 little girls applauded—one of them could even play Romanian Dance No.1. There was tea afterwards, with the headmistress presiding—but still I don't know just how much I'm to get for this excursion: £10 or 15?!)[33]

Two days later Bartók again went on a day trip to the country, for a public afternoon performance in Bournemouth probably arranged by his London concert agent. It was a significant day in British politics, for a General Election was in progress which would result in the first Labour-dominated government in the country's history. In Bournemouth Bartók performed in the role of orchestral soloist for the first time in Britain since 1905, playing his *Rhapsody*, Op. 1 with one of the country's finest regional orchestras, the Bournemouth Symphony. Sir Dan Godfrey, its conductor, was by then in his thirty-first year of association with the orchestra. He prided himself on the careful rehearsals given to new works and on a wide repertory which made the most of the slender thirty-nine member orchestra.[34] Concerts were held in the Winter Gardens, a draughty edifice of glass and iron with dubious acoustics. Despite the cold weather, substandard seating, and the recent establishment of a local radio station, the Thursday afternoon series at which Bartók performed was still a social high point of the week and was well attended.[35]

Reviews of the concert were generally inept, confusing Bartók's more recent compositional ideals with his intentions at the time of writing the *Rhapsody* nearly twenty years before. Much of the blame lay with the distributed programme-notes, in which this confusion was persistent.[36] None the less, because of its late-Romantic outlook the work gained favour with a majority of the critics, as did three short Bartók solos presented later in the afternoon. The critics from the *Bournemouth Guardian* and the *Bournemouth Graphic* both appreciated the melodic and rhythmic features of the *Rhapsody*, and found Bartók's interpretations of his works attractive.[37] In the *Bournemouth Times*, however, the anonymous critic was

[33] Letter, Bartók to his mother, 4 Dec. 1923, *BBcl* pp. 345–6.
[34] Dan Godfrey, *Memories and Music* (London, 1924), pp. 284–5.
[35] Gordon Bryant, 'Bournemouth Winter Gardens', *MNH* 19 Jan. 1924, pp. 64–5.
[36] MTA BH 2049/124. The 'Analytical and Historical Notes' were probably written by Hamilton Law.
[37] Reviews mentioned: Anon., 'Bournemouth Symphony Concerts', *Bournemouth Guardian*, 8 Dec. 1923; 'Cecilia', 'Music of the Week', *Bournemouth Graphic*, 14 Dec. 1923; Anon., 'Music in Bournemouth', *Bournemouth Times*, 8 Dec. 1923.

less accepting of Bartók's music and expressed a wish for the traditionally accepted qualities of Hungarian music:

We found little or no semblance to anything beautiful in the way of thematic material employed in this rhapsody, which was the very feature [in which] one would expect it to be most enriched. What was most disappointing was the poverty of the thematic material which failed to bear out the reputation usually associated with Hungarian music. This may be a source of congratulation, or at least a tribute to the native excellence of our own music. It is not difficult to think of a dozen or so rhapsodies based on English, Scotch, Irish and Welsh melodies which for beauty of their indigenous tunes are incomparably superior.

As for Bartók, he enjoyed the day by the sea and had no complaints about the concert. On the reverse side of a picture postcard showing the Bournemouth Sands he wrote to his son: 'Hello Béla, my boy. This afternoon I walked along the coast (overleaf) and gathered up all kinds of seaweed. I'll bring them home. The concert was this afternoon; the orchestra behaved well enough.'[38]

The final girls' school recital took place on 12 December at St James's in West Malvern, probably in collaboration with the nearby Abbey School. On an overcast, wintry day Bartók journeyed from London to present the same programme as at Aldeburgh, except that he decided to add, at the start of the second bracket, nine short pieces from his *For Children* collection, probably as a concert demonstration of works being studied by some of the students.[39] Bartók was introduced to the audience by Margaret Woodhouse, who taught music at the school. A summary of her talk was recorded in the school magazine *St James's Gazette*:

M. Béla Bartók's own compositions, as Mrs. Woodhouse quoted from a London paper, appeal 'only to the connoisseur'. Yet his recitals have been cordially received in many schools as evidenced by another London paper, which states: 'The younger generation have [sic] not appeared in the least troubled by his atonality, whilst attracted by the simplicity and humour particularly of his children's pieces.'[40]

No more is known about this concert. After it was over Bartók hastily returned to London and left the following day for Paris, where more important concerts took all his attention. In response to the concert, however, George Woodhouse penned a further article about Bartók for The Abbey's magazine the *Hazel Nut*.[41] Woodhouse let his thoughts wander, recording frankly the questions in his mind and some half-formed answers—not the sort of thing he would have allowed to be published in a serious musical journal. He suggested that lay musicians might more readily

---

[38] Postcard, Bartók to his son Béla, 6 Dec. 1923, *BBcl* p. 346.
[39] MTA BH 2049/132.
[40] Anon., 'Music Notes', *St James's Gazette*, Summer and Autumn Term 1923.
[41] Béla Bartók', *Hazel Nut*, Christmas Term 1923.

adjust their 'mental lens' to appreciate Bartók's music than professionals, who were often biased by concepts and theories imposed by older schools of thought. He confessed a wish for an easy way of understanding Bartók's music, but admitted that he had not found one. What music he did understand, however, convinced him that Bartók was a composer of 'extraordinarily concentrated and strong' musical thought, a quality which would inspire an enduring, if not large, following. Looking to the future, Woodhouse concluded: 'We would gladly welcome the genius who would interpret the spirit of a new age in music. We have the talents and the artists, but have we the genius who is also the man? Bartók stands almost alone in the new modern school of composers to inspire an optimistic answer.'

Between the various provincial engagements of his two British visits of 1923, Bartók took part in at least seven performances in London, most in collaboration with members of the Arányi family. Only one was a fully public affair. This concert was held on 30 November at the Aeolian Hall, as part of a series of concerts organized by Philip Heseltine's friend, Gerald Cooper. Bartók and Jelly Arányi performed a programme in which the chief items were Beethoven's Violin Sonata in G major, Op. 96 and Bartók's Second Violin Sonata (1922).[42] The other London concerts were given before musical associations or private audiences and were most varied: from the evening with the International Society for Contemporary Music on 7 May, at which both Bartók's violin sonatas were played, to a session of domestic music-making with the Fachiris at their Chelsea home.[43]

In contrast with the scant coverage of provincial concerts, Bartók's more substantial metropolitan appearances elicited an abundance of critical opinions: lengthy and laconic, friendly and ferocious. Through these writings it became clear that overall Bartók was losing rather than gaining adherents. It was now easier than in 1922 for critics and the public to reach a decision about his artistic worth, as he only presented a limited range of works and had no compunction about repeating items frequently. By the end of 1923 some Londoners had heard his First Violin Sonata five times and naturally knew its contours well. Despite significant new items in some 1923 programmes, a number of his piano pieces—*Allegro barbaro*, 'Evening in the Country', and 'Bear Dance'—were even better known. This familiarity helped wavering critics and listeners to make up their minds. On balance, they decided negatively. As a *Daily Telegraph* critic, probably Robin H. Legge, concluded after rehearing the Second Violin Sonata:

This second hearing decreased the difficulties of judgment, but increased the dislike which had been indefinitely formed on the other occasion. The dislike is engendered, not, indeed, by any heresy in Bartók's methods, but by the monotonous orthodoxy

[42] MTA BH 55 (preliminary), MTA BH 2049/129 (final).
[43] See, respectively, MTA BH 2049/116, and MTA BH 2049/130. Details of all the London concerts of 1923 are provided in notes 11 and 12, above.

with which he follows his lines of thought. There are but few lines which he follows, and all are so engrossedly introspective that those who would go after him must needs yield up everything to the pursuit.[44]

Not all critics hardened in their judgement of Bartók, but the largest body, led by the *Observer*'s Percy Scholes, showed unmistakable signs of hostility. Significant among the 'defectors' to the negative side during this year were two of Bartók's earliest British supporters, Leigh Henry and Cecil Gray. Both had become convinced that the dissonance which permeated Bartók's latest works indicated his artistic regress.[45] Their more senior colleagues, notably Ernest Newman and Edwin Evans, were also worried about the direction of Bartók's recent works, but displayed more caution in their commentaries. They took care to point out successful aspects of performance and to stress the sincerity of utterance in Bartók's compositions. Newman, indeed, heard somewhat less ugliness and more compositional ingenuity as he became better acquainted with Bartók's music.

The frequent repetition of items also encouraged the critics to turn from the construction of works to their rendition, and an increasing number came to side with Percy Scholes in seeing Bartók as a 'jerkily rhythmic automaton'.[46] Between Bartók and these critics there lay a vast conceptual gap. Brought up with respect for the Matthay school of playing, with its emphasis on relaxation and careful use of weight,[47] many viewed Bartók's avowedly percussive approach with horror. While their 'common sense' dictated that a piano key should be 'depressed', Bartók believed it should be 'struck', in the interest of producing a clear, sharp edge to a note. After Bartók's second tour of the year Richard Capell felt compelled to comment: 'Mr. Bartók's piano playing is extraordinarily unpleasant—heavy as lead. It would be interesting to hear his piano pieces from someone with a different sense of touch.'[48] Capell could not dismiss this unpleasantness on the simple grounds of poor taste or inappropriate choice of finger technique. To his mind Bartók's musicianship had to be questioned: 'the composer's piano-playing is so horrid. . . . Mr. Bartók persistently shows us the worst side of the piano, and that his insensitive touch is his misfortune and not his intention was proved by his cruelty to Beethoven's lovely sonata, Op. 96.'[49] Not everyone accepted Capell's opinion, of course. Ernest Newman, for one, deemed this same Beethoven rendition 'extremely intimate'.[50] Another

---

[44] 'London Concerts', *DT* 3 Dec. 1923.
[45] Leigh Henry, 'London Letter', *Chesterian*, NS no. 36 (Jan. 1924), 153; Gray's changes in opinion are documented extensively in Chapter 7, below.
[46] As expressed in 'Music of the Week', *Observer*, 26 Mar. 1922.
[47] For a statement of Tobias Matthay's views see *The Act of Touch in all its Diversity* (London, 1903).
[48] R. C., 'In the Concert Room', *MMR* 54 (1924), 13.
[49] R. C., 'Music Notes', *DM* 3 Dec. 1923.
[50] 'The Week's Music', *ST* 2 Dec. 1923.

critic, from the *Lady*, found Bartók's very austerity attractive: 'He [Bartók] makes no effort to charm us into liking his music by variations of touch and tone—Yet in his undecorated, massive, and austere playing there is something noble and simple which appeals to me.'[51]

As the critics formed more definite opinions the language as well as the content of their columns became more varied. In some cases a high-flown tone was merely a mask, disguising the critic's continuing problems in grappling with the music. This was certainly the case for the *Westminster Gazette*'s critic, who referred in one column to 'the futuristic, not to say Bolshevistic, enfant terrible, Béla Bartók'.[52] More sensational, even cruel, was the way Percy Scholes described his feelings at Bartók's ISCM concert: 'I suffered more than upon any other [occasion] in my life—apart from an occasional incident or two connected with "painless dentistry". To begin with, there was Mr. Bartók's piano touch . . . "He had a touch like a paving stone".'[53] The opposite extreme, of polite, generous evaluation, was seen only in the writings of Calvocoressi, for whom Bartók remained 'one of the leading forces among today's composers'.[54] The heat and vituperation generated by Bartók's performances in 1923, particularly his public concert on 30 November, so disquieted Calvocoressi that he decided to issue a public warning to his colleagues:

Generally speaking, nothing could be more instructive to both readers and critics than such comparisons: they remind the former that judgments may differ widely pending the time when more careful, more sober thinking, and the interplay of conflicting views, will have ensured the survival of the fittest. And to the latter they come, like the mummy at the feast, to utter the warning: 'Be merry while you may, but remember that there is a hereafter.' In the present instance, the range between extremes, of opinion is somewhat wider than is usual. This, I believe, is enough to show how needful it is to study Bartók's music carefully and dispassionately. For never has it occurred that a composer in whose wake so marked a conflict of opinion could arise was ultimately judged unworthy of admiration.[55]

In no small measure Bartók's problems of reception in 1923 stemmed from the two works which featured so significantly in these London programmes: his violin sonatas. He thereby tackled his audiences with two of the most extreme works that he would ever write, and drew attention to a style from which he was already starting to retreat. It was not surprising, therefore, that in their more detailed assessments the critics wrote of excesses, ugliness, inaccessibility, and incoherence. Of the two sonatas the second was generally found to be the less palatable, perhaps because of its lack of the traditional three movements. To Herbert Antcliffe this work

[51] C. M., 'Musical Notes', *Lady*, 17 May 1923, p. 536.
[52] Anon., 'Béla Bartók', *Westminster Gazette*, 1 Dec. 1923.
[53] 'Music of the Week', *Observer*, 13 May 1923.
[54] 'The Newest Music from Abroad', *Outlook*, 22 Dec. 1923, pp. 469–70.
[55] 'Bartók's "Duke Blue Beard Castle" ', *MMR* 54 (1924), 36.

sounded more like a series of disjointed sketches than a homogeneous work.[56] Any beauty in it was found in individual sections rather than in the whole. After a second hearing Richard Capell still found the second sonata a challenge and suggested that it represented 'the battle-cry of Eastern barbarism at our gates'.[57] He looked on the work as a new style of Hungarian Rhapsody, with the slow first movement expressing 'a melancholy too outlandish for us to make anything of it', and the second indicating something of 'impatience, picturesque invective, caprice and animal energy'. Even friendly 'moderates' among the critics were hard put to find many encouraging words about the Second Violin Sonata. On first hearing the work Edwin Evans wrote of 'purposeful expressionism', but then admitted that he did not quite grasp that purpose himself![58] Hearing it again in November he was still at a loss, although most politely so: 'The second of Bartók's two Sonatas is more concise than the first, but for that reason, also less self-explanatory and perhaps less accessible—unless we are prepared to meet the composer more than half way.'[59] Ernest Newman was probably in America when this sonata was first performed in London, but heard the second performance at the Aeolian Hall: 'It is the genuine expression of a genuine train of thought and a genuine point of view', he boldly stated but then confessed that he could not take such a work to heart.[60] As so often during these early visits to Britain, Percy Scholes outdid his colleagues in disparaging this sonata. After a second hearing he felt confirmed in his opinion that the music was 'weak' and 'childish':

There is nothing 'to it'. It has no beauty and no strength, it expresses nothing, and it is not even, in any real sense, 'modern', much of it being merely a vague wandering about one or other of the twelve dear old major keys with an obtruding extra note or two in every chord to vary the landscape.[61]

Scholes's opinion was indeed extreme, but it was probably the most widely disseminated in Britain. He not only wrote for the *Observer*, but since February 1923 had been the first music critic for the British Broadcasting Company. Thereby, he transmitted his opinions on 'The Week's Music' to nearly 600,000 wireless sets throughout the country.[62] Although Scholes saw radio as 'the greatest boon to music since Jubal struck his lyre',[63] his racy commentaries were no boon to Bartók.

[56] 'Bartók, Brahms, Borwick and Some Others', *MNH* 19 May 1923, pp. 492–3.
[57] *DM* 3 Dec. 1923.
[58] E. E., 'London Concerts', *MT* 64 (1923), 424.
[59] E. E., 'Jelly d'Arányi and Bartók', *MT* 65 (1924), 71.
[60] *ST* 2 Dec. 1923.
[61] 'Music of the Week', *Observer*, 2 Dec. 1923.
[62] It is probable that Scholes included the above review in his weekly broadcast at 7.10 p.m. on 6 December. The programme was broadcast over stations in London, Birmingham, Bournemouth, Manchester, Glasgow, Cardiff, and Newcastle.
[63] Maurice Gorham, *Broadcasting and Television since 1900* (London, 1952), 47.

With such a formidable new violin work demanding assessment, Bartók's solo piano pieces came in for less comment than in the previous year. The opinion was widely expressed, however, that these short piano works were most successful when percussive means were used to gain a lyrical effect, as in the *Bagatelles*, Op. 6 and the various folk-music arrangements.[64] When Bartók tried to be lyrical by the use of more conventional techniques, on the other hand, the results were considered disappointing. The *Dirges* Op. 9a were cited in this regard.[65] As in 1922, Bartók's selections from *For Children*, which displayed his techniques most simply, did inspire some critics to write rapturously. For the *Daily Telegraph*'s critic the eight pieces based on Slovakian folk-tunes which Bartók played on 11 May were 'in a class by themselves, completely delightful and wholesome, and for these some of us would forgo some of his more serious compositions'.[66]

In his London performances of 1923 Bartók managed to convey an impression of grim, tight, relentless activity, both physical and mental. 'His music is rather like a volcanic deposit', concluded the *Lady*'s writer.[67] 'The Bartók system of composition and performance is one of the most rigid-minded-rigid-muscled ever invented', wrote Percy Scholes; 'in shunning sentiment Bartók has lost beauty, in shunning rhetoric he has lost reason.'[68] The words of the *Daily Telegraph*'s critic well summed up popular opinion:

The truth is that with Bartók there is no incentive given to attract us to take this uneventful road of atonality, and this because of his unfortunate serious-mindedness. His weakness is not because of ugliness, but because of scepticism; like Sir Andrew Aguecheek, he cares not for the songs of good life.[69]

The feeling of satiation with Bartók's limited and unattractive repertory was heightened by the presentation of a variety of musical novelties in London during the last two months of 1923. Strauss's 'Alpine' Symphony, Op. 64 finally received its first London performance on 13 November. Schoenberg's *Pierrot lunaire*, Op. 21 was likewise premièred on 19–20 November in a blaze of publicity, with Darius Milhaud conducting the work at three different London locations. By this later event, in particular, the critics were mightily roused. 'Ugly and pointless', wrote Ernest Newman.[70] 'Herr Schönberg is not deaf, but when I listen to his music I often begin to think that I am', commented none other than the President of the ISCM, Edward J. Dent.[71] This controversy was unfortunate for Bartók.

[64] See, e.g., E. E., 'Jelly d'Arányi and Bartók', *MT* 65 (1924), 71.
[65] See, e.g., ibid., and Anon., 'London Concerts', *DT* 14 May 1923.
[66] *DT*, 14 May 1923.
[67] C. M., 'Musical Notes', *Lady*, 6 Dec. 1923, p. 608.
[68] *Observer*, 13 May 1923.
[69] Anon. [probably Robin H. Legge], 'London Concerts: Béla Bartók', *DT* 3 Dec. 1923. Sir Andrew Aguecheek is a character in Shakespeare's *Twelfth Night*.
[70] 'The Week's Music', *ST* 25 Nov. 1923.
[71] 'The Co-Pessimists', *The Nation and the Athenaeum*, 1 Dec. 1923, p. 358.

Many of the critics, drawn by the superficial similarities of style, were all too prepared to 'tar him with the same brush'. But harsher treatment would follow. Other pianists of fame were in town, including Dohnányi and Henry Cowell, who sometimes played with his elbows and plucked piano strings.[72] Cowell's London recital fell on the same day as Bartók's concert at the Woodhouse Pianoforte School. The critics flocked to experience the new sensation from America and to write mocking reviews for their editors. Who wanted to write about yet another repetitious Bartók performance after such an exhibition? Only loyal Calvocoressi recorded what would be Bartók's last performance in London for four years.

In 1923 the critics struck back. There was a clarification and entrenchment of opinion, to Bartók's long-term cost, as many of the more hostile lines of argument would endure for years as standard responses to his music. The optimism generated in 1922 had begun to sour in Bartók's mind: Britain was no more inclined than the other countries of Europe to afford an easy recognition to his talents. In February 1925 he would write to Calvocoressi: 'The connections between England and me seem totally to have been broken off; neither am I invited nor are my works played there. It is partly my own fault, since I have become an ex-composer.'[73]

---

[72] Bartók and Cowell were both house-guests of the Wilsons at this time. In a letter to Halsey Stevens, of 10 Oct. 1950, Cowell recalled Bartók's interest in his tone-cluster technique, and Bartók's letter early in the following year asking for 'permission' to use this technique in his own works. (Halsey Stevens, *The Life and Music of Béla Bartók* (2nd edn. New York, 1964), 67.)

[73] Letter, Bartók to Calvocoressi, 15 Feb. 1925, in German, in BBrCal pp. 221–4.

5.

BARTÓK OVER BRITAIN

'Music is the common property and common enjoyment of mankind. It is one of his rare delights.'[1] So wrote the 37-year-old Managing Director of the British Broadcasting Company in *Broadcast Over Britain*, his 1924 blueprint for the future of British wireless transmission. When appointed in December 1922, John Reith came to the helm of a minute organization with less than three months of broadcasting experience. Ten years later he would be considered by many to be the most powerful man in Britain, except for the Prime Minister.[2] By that time half of Britain's families would own a wireless—a wireless which only provided BBC programmes. In contrast with their American counterparts, British authorities had from the start considered Marconi's invention too precious—in the wrong hands, too dangerous—to be entrusted to the whims of a free market. Radio was to be used for the highest possible goal, the enlightenment of man, and never for his degradation. Within appropriate limits it could entertain, but it must also educate.

The coming of radio placed the musical destiny of the country, to large measure, in the hands of the BBC. The apparatus enabled millions to hear music previously reserved for a privileged few, and through its programmes and commentaries was able to manipulate public taste and opinion. Reith, in 1924, had stated a gloriously democratic aim for the organization's music programmes, but the practical realities were less august. From its inception the BBC could only hope 'to please 75 per cent of the listeners 75 per cent of the time'.[3] Even that level of compromise was rather optimistic, as Joseph Lewis, founding Musical Director at the Birmingham studios, explained in 1924:

A wants Stravinsky, B requires Wagner, C only 'reacts' to Russian music, D makes a 'gesture' towards musical comedy, E craves for Gilbert and Sullivan Opera ... F must have jazz and red-nosed comedian stuff, and so on ... And at the present time my opinion is that the only music which transmits really well is the 'linear' type of music such as Haydn, Mozart, Beethoven. Still, later we may be able to give A, B, and C their hearts' (or ears') delight, but when it comes to category F, it is a horse of a different colour. Under no circumstances will the B.B.C. ever pander to low tastes, and the best thing for a listener who requires a sort of music-hall annexe to his home, is to chop down his aerial and play Mah Jongg instead.[4]

[1] J. C. W. Reith, *Broadcast Over Britain* (London, 1924), 173.
[2] See Ronald Blythe, *The Age of Illusion* (London, 1963; rpt. 1983), 50.
[3] Sir Henry Hadow *et al.*, *New Ventures in Broadcasting* (London, 1928), 63.
[4] 'Broadcasting Music', *Sackbut*, 4 (1923–4), 227–8.

Bartók's music clearly fell into Category A, but he soon gained a hearing. On 6 November 1924 his *Romanian Folk Dances* (1915) were broadcast by the BBC, and every year since then at least one of his works has been included in its programmes.

Coupled with the advent of 'talking films' in the late twenties, radio caused important changes in the profile of British music-making, both directly and indirectly. The new, democratic access to music hastened the decline of private concert activity and brought many concert-halls into financial trouble. These difficulties were partly the result of a false expectation that the musical boom of the early 1920s would continue unabated,[5] but also stemmed from the fact that in its early days radio was undoubtedly drawing away potential concert-goers, who were anxious to get the most value out of their radio licences. In 1926 the Queen's Hall, London's première orchestral venue, showed signs of failing financially. Sir Henry Wood's Promenade series, which had survived the war, now seemed headed for collapse.[6] While some orchestras, such as the Hallé, sought to form alliances of 'live performers' to withstand the onslaught of radio, others, including Wood's orchestra, desired a collaboration with broadcasting.[7] In the long run this latter partnership was more productive, as radio introduced new forms of music to listeners who then often sought to attend live performances. Several years after the crisis Sir Harry Brittain observed:

Many of these young people, who stand about uncomfortably for hours listening to modern compositions by Stravinsky and Ravel, have been musically educated by the B.B.C. And although the programmes of the Promenade Concerts are broadcast into millions of homes, thousands of men and women attend them at the Queen's Hall. And so it is with other concerts. The musical taste of the nation was never more sound and democratic than it is to-day.[8]

Among performers there was naturally a scramble for radio engagements. They provided an advertisement, often nationally, of a player's or ensemble's talents and resulted in a lucrative and predetermined fee which bore no immediate correlation with audience size. The attraction to Bartók of radio performance was immense, and in Britain he sought BBC engagements to the exclusion, for some years, of all other forms of music-making. When for two short appearances he could gain at least £60, there was little point in undergoing the hassle and anxiety of arranging public concerts.

The first mention of a BBC engagement for Bartók came from a former

[5] See G. A. Pfister, 'Empty Concert Halls', *MNH* 12 Apr. 1924, pp. 352–3; Edwin Evans, 'Half-Time in England', *Modern Music*, 3. 4 (May 1926), 10–15.
[6] See David Cox, *The Henry Wood Proms* (London, 1980).
[7] See Anon. [Basil Maine], Editorial, 'The Queen's Hall Crisis', *MB* 8 (1926), 319–20.
[8] *The ABC of the B.B.C.* (London, n.d. [1932]), 53.

pupil Dezső Rácz, then Information Officer at the Hungarian Consulate in
London. In a letter of 3 December 1925 he encouraged Bartók to give in
Britain the first performance of the piano concerto which he was then
planning to write, possibly as the 'sensational' centrepiece of an all-Bartók
orchestral programme.[9] Three months later, Rácz's vague enthusiasms had
taken firmer form.[10] There was the possibility of an evening of Hungarian
music at the BBC, as part of a series of concerts devoted to recent trends in
European music. Rácz reported to the Hungarian Foreign Ministry: 'An
evening broadcast means ten million listeners and could provide un-
imaginable publicity for Hungarian music.' Bartók was interested in the
idea, and so over the next year plans were slowly drawn up. In preparing for
the evening Rácz gained practical help from Hubert J. Foss, at Oxford
University Press, Frank Whitaker, Editor of the *Star*, and his own Foreign
Ministry.[11] The concert was to be performed in London, starting at 9.35
p.m. on 15 March 1927, and to be broadcast from the BBC's radio station
at Daventry, 5XX.[12] Highlights of the announced programme were
Bartók's new Sonata (1926) for piano and his *Five Songs*, Op. 16, sung by
the Hungarian soprano Mária Basilides. With great care Rácz smoothed the
way for his old teacher, negotiating a fee of £25, attending to all personal
details of his trip, and arranging for various articles and photographs to
appear in the press.[13] When all appeared settled Rácz discovered from a
statement in the Hungarian press that Bartók was ill. Anxiously, on 7
March, he wrote to him, outlining the best course of action if the concert
had to be postponed and begging for an early indication from Bartók of his
intentions.[14] The following day he wrote again, clearly worried that this
enterprise which he had nurtured for so long might fail and the considerable
promotion costs be to no purpose.[15] Bartók sent back the bad news: he was
too ill to come. The BBC, keen to reschedule the concert later in the year,
despatched the polite telegram: 'REGRET YOUR ILLNESS AND POSTPONEMENT
CONCERT WISH SPEEDY RECOVERY = BROADCASTING +'.[16] In several British
newspapers Bartók's inability to appear was noted with regret.[17] He had
gained considerably in reputation since his last visit in 1923 because of the
popularity of his *Romanian Folk Dances* (1915) and the *Dance Suite*

[9] MTA BA-B 3477/160.
[10] Report, Dezső Rácz to the Hungarian Foreign Ministry, 6 March 1926, partially
reproduced in Arisztid Valkó, 'Bartók Béla tervezett londoni hangversenyének levéltári
háttere', *Magyar Zene*, 18 (1977), 100.
[11] For documents relating to the arrangements for this concert, see: BL Add. MS 51023B;
Országos Levéltár (Budapest), Foreign Ministry reports, in Arisztid Valkó, 'Bartók Béla',
pp. 99–105; MTA BH 2095 and 2096/a.   [12] MTA BH 2096/b.
[13] See, respectively: MTA BH 2095; Anon., 'Béla Bartók to Broadcast', *ES* (London), 7 Feb.
1927, and Anon., 'Bártok-hangverseny Londonban', *Budapesti Hírlap*, 13 Feb. 1927.
[14] MTA BH 2096/a.
[15] MTA BH 2097; Arisztid Valkó, 'Bartók Béla', p. 104.   [16] MTA BH 216.
[17] Anon., 'Bartók Béla lemondta londoni rádióhangversenyét', *Budapesti Hírlap*, 20 Mar.
1927. These London papers included *The Times* and the *Morning Post*.

(1923). Many music-goers were now curious to hear his most recent works to see if the more popular style of these known compositions had been continued. Even despite Bartók's non-attendance the purpose of the evening was not completely lost, for Mária Basilides was still engaged for the concert and did fulfil her part of the programme.[18]

Rácz was not content to let the BBC engagement be forgotten, already on 14 March writing to Bartók with new plans.[19] He suggested performances of chamber music with Jelly Arányi or the cellist Beatrice Harrison at the end of June, when the concert and social season was still at a high level. When Bartók finally made his first appearances with the BBC in October, however, the central feature was the work which Rácz had requested nearly two years before, the First Piano Concerto (1926), although not now as a world première. The arrangement with the BBC for these October 1927 concerts established a pattern for several later visits. For a 'package fee' Bartók gave a lighter studio performance as a soloist or chamber player, and also performed in a more substantial role, as orchestral soloist or performer in a lengthy chamber concert. On 9 and 10 October 1927 his all-inclusive fee was 55 guineas.[20] In the earlier concert, starting at 9 p.m. on the Daventry station 5GB, he joined with well-known London musicians in a performance of Mozart's Piano Quintet in E flat major, K 452, followed by two solo brackets, in one playing Italian cembalo music transcriptions, in

'A Famous Modern Composer' (Radio Times, 7 October 1927)

[18] MTA BH 2099.   [19] DB.iii pp. 132–4.   [20] MTA BH D-II.28 Számlák.

the other the more popular of his piano pieces. On the following evening, at the prime time of 8 o'clock, Bartók gained a wider coverage on London 2LO and Daventry 5XX stations through an hour-long programme devoted exclusively to his orchestral music. The Wireless Symphony Orchestra collaborated with Bartók in a performance of his new piano concerto as well as *Two Portraits*, Op. 5 and his popular *Dance Suite* (1923).

Although Bartók broadcast to millions of listeners in his many BBC concerts, few traces of these performances or their public reception remain. The BBC did not record them,[21] and unless 'outside broadcasts' from one of the London concert-halls were involved, they were infrequently reviewed in the music or broadcasting columns of the press. Bartók's first performance on British radio, therefore, received no special mention in the press. His second, however, did, for the rendition of the First Piano Concerto engendered a storm of comment across the country. Writing in 1931, Arthur G. Browne recalled:

... Bartók's music did not reach the general public until nearly four years ago, when the British Broadcasting Corporation performed his Piano Concerto. This work created something of a sensation. With commendable courage, the B.B.C. continued to broadcast Bartók's music in the face of that peculiarly intense criticism with which 'modern' art is greeted in this country.[22]

The problem was a simple one. 'An Englishman's home is his castle', and he resented being invaded by these alien sounds. Cecil Lewis, a broadcasting reviewer for the *Observer*, accused the BBC of starting a 'Black Monday series'. He found Bartók's music either 'laboriously amorphous or maddeningly repetitive'.[23] From the performance, Lewis concluded, many would certainly remember the name of Bartók: 'He must have caused a series of aesthetic earthquakes all over the British Isles and brought home to many the truth that the greatest boon about wireless is the ability to switch it off.' Many scandalized listeners wrote to journals, newspapers, and the BBC with their impressions of Bartók's music. In the *Musical Standard*, J. Alfred Johnstone's analysis appeared:

I should suggest that it seemed to me something like a band of nocturnal cats ('voluptuous caterwauling') mingled with an occasional braying of an ass (bassoon), and perhaps pervaded by the instrumental efforts of a company of lunatics into whose ignorant hands a set of orchestral instruments was placed, with lurid instructions to 'do their worst' ... I think he was trying to give a vivid portrayal in sounds of the ideas of insanity, Bolshevism and iconoclasm, picturing them to himself as spirits wandering uneasily over a world wrecked into chaos.[24]

---

[21] This has been confirmed after a thorough search of the BBC's Sound Archives (unpublished letter, BBC Secretariat to the author, 27 Mar. 1984).
[22] 'Béla Bartók', ML 12 (1931), 35.
[23] 'Broadcasting', *Observer*, 16 Oct. 1927.   [24] MS 5 Nov. 1927, p. 165.

Among more liberal-minded music-lovers the judgements on Bartók's concert were not as damning. The journal *Musical Opinion* featured Bartók extensively, with a biographical article, photograph, and long review.[25] His effect had been volcanic, the review asserted, but 'we only wish for more after this experience'. With the first of the *Two Portraits* Bartók's genius was immediately apparent, through its use of colour, its clear polyphony, and the concentration of its purpose; in short, it was 'a picture of stainless beauty'. By contrast, the second *Portrait* was 'impish' and 'bizarre', almost Straussian. The Piano Concerto which followed presented the reviewer with more problems than the other two works on the programme, although he approved of the pliant use of national material and the relationship between soloist and orchestra. Such a work, none the less, did demand 'generous concessions from those whose outlook has been moulded upon familiar patterns', especially when the work was presented with insufficient rehearsal. The performance of the concerto had, indeed, almost been cancelled because of the late arrival of parts.[26] For its première at Frankfurt-am-Main under Furtwängler there had been six rehearsals; planned performances in New York later in the year had to be cancelled because the conductor Willem Mengelberg and his orchestra found the work unplayable. In London, with only one rehearsal, a coherent if not polished performance had been given, a credit to the British strength in sight-reading and the experience of the conductor Edward Clark. A student of Schoenberg before the war, Clark had been Musical Director at the BBC station in Newcastle-upon-Tyne until 1926, when he was moved to London as a programme planner.[27] There, for the next decade, he sought to promote contemporary music, in the process establishing and furthering contacts with nearly all the leading composers of the day.[28]

This concert was performed before a sizeable audience, as colourfully described in the Hungarian-language Czechoslovak paper *Prágai Magyar Hírlap*:

London, 14 October. Béla Bartók, the world-renowned Hungarian pianist, recently appeared as a guest artist in the studio of the English radio. The entire English press describes Bartók's concert as an artistic happening quite out of the ordinary. The huge significance of the Hungarian artist's concert is attested by the fact that, quite against the rules of the radio company, a large number of listeners got into the studio. Because of their names and their exalted positions in musical circles the

[25] Anon., 'Personal and Otherwise: Bartók in Person', *MO* 51 (1927–8), 147 (with accompanying photograph); Anon., 'The Concert Season: Béla Bartók', ibid. 155.
[26] *BBcl* p. 416.
[27] For details of Clark's work with the BBC see Elisabeth Lutyens, *A Goldfish Bowl* (London, 1972), 115–36.
[28] See Anon., 'Personal and Otherwise: Mr. Edward Clark', *MO* 51 (1927–8) 148 (with accompanying photograph). A collection of Clark's correspondence with leading composers is found in BL Add. MS 52256.

studio management had to let them in. The listeners included—among others—Gustav Holst, England's most celebrated composer, the son of Lord Oxford [*sic*], the well-known music critics Eaglefield Hull and Whitaker, the writer Rosa Newmarch, the lady-pianist Woodhouse, and many others. The Bartók celebration lasted late into the night.[29]

With this attention, the troubles taken by the BBC, and the good fee, Bartók was well pleased with his three-day stay in Britain. Although he had already performed on radio a number of times in other countries, he realised that these most recent broadcasts were a milestone in his career, establishing contacts which would be vital for the success of his future visits to the country. The broadcasts were exciting because they provided the first opportunity ever for him to play with an orchestra in London, and also because these performances could be shared across the airwaves with friends and relatives back home. Bartók's mother had twice been primed to listen in at the correct time![30]

Although less starkly dissonant than his violin sonatas, the concerto with its primitive melodies and innovations in rhythm and orchestral colour was in no way conventional. The work was, therefore, most effective in awakening a mass, nation-wide consciousness of Bartók's music and ideals. This interest was sufficient to maintain itself in the months after his visit, and even to generate some fierce debates. There were frequent reports in the British press of his many concerts in the United States during the following months. On 14 December the BBC gave a repeat performance of the *Romanian Folk Dances*. In provincial cities, too, Bartók's music was being heard. During February 1928 some of his works were presented at the Bournemouth Centre of the British Music Society; in March one of his quartets was played at the Manchester Contemporary Music Centre; in April the American pianist Keith Corelli included some Bartók pieces in a Liverpool recital.[31] Meanwhile, the *Musical Times*'s reviewer of piano music systematically reviewed the available Bartók piano literature.[32] London audiences were, furthermore, introduced to Bartók's *Two Pictures*, Op. 10 during the visit of the Budapest Philharmonic Orchestra in June 1928. The reviews were surprisingly full of praise for Bartók's pre-war composition. *The Times*'s writer even complained that these 'blameless and charming pieces . . . hardly displayed him as an effective champion of modernity'.[33]

At the BBC a 'Bartók debate' was in full swing. Having received hundreds of letters after Bartók's two broadcasts, Percy Scholes penned a leading

---

[29] Anon., Bartók Béla meleg ünneplése Londonban', *Prágai Magyar Hírlap* (Prague), 15 Oct. 1927. Lord Berners was probably meant by 'son of Lord Oxford'. The pianist mentioned was probably Violet Gordon Woodhouse, rather than Margaret (Mrs George) Woodhouse.

[30] *BBcl* pp. 414, 416.

[31] See, respectively: *MB* 10 (1928), 117; ibid. 80; MTA BH 2049/233.

[32] See T. A., 'New Music: Pianoforte' columns, *MT* 68 (1927), 432; 69 (1928), 36, 427, 997.

[33] Anon., 'Budapest Orchestra', *The Times*, 19 June 1928.

'Is Bartók Mad—Or Are We?' (Radio Times, 9 December 1927)

article in the *Radio Times* entitled 'Is Bartók Mad—Or Are We?'[34] His initial answer to the question was 'neither'. It was natural, he argued, for a composer to wish to write in a new idiom, and equally natural for the listener's ear to find difficulty for some time in accepting these new sounds. Coming from one of Bartók's most vocal critics, Scholes's reasonableness of argument must have surprised many of his readers. Indeed, the recantation of his previous beliefs about Bartók was to follow:

I have just been turning up criticisms of his music I wrote when I was music critic of the *Observer*. I hope the poor fellow didn't read them! For I see that after a period of doubt I definitely pronounced him *no composer*. I believe the compositions I heard in those days were very immature, and perhaps that misled me. If I can find excuses for myself I will. I am very human! And now I come to the point. Candidly and sincerely, I assure you that the result of certain recent broadcast programmes has been to make me believe I was formerly in error about Bartók. I am not very sure, but I *believe* that Bartók is a great composer. I am very sure he is a clever one. I have a suspicion that what has been wrong with listeners who have written to protest against Bartók's music has been, in the main, not Bartók's composing but their hearing. *The human ear is a very conservative member.*

Many of Scholes's readers were unconvinced. From Gloucestershire one wrote:

His [Scholes's] half-hearted attempt to prove that beauty is in these days suspect, and that we *ought* to like ugliness, hardly convinces himself. It won't do. I agree with him that musical tastes change from period to period, but Stravinsky's and Bartók's stuff isn't *music* at all! And to talk of it as any form of that divine art is an outrage.[35]

[34] *RT* 9 Dec. 1927, pp. 525–6.
[35] T. F., 'Good Lord, Deliver Us!', *RT* 6 Jan. 1928, p. 3.

From Nottingham came a more laconic response: 'It may be modern, it may be music, it may appeal to savages, but to the lovers of the beautiful in music it will never appeal.'[36] Scholes's view did, however, find some support. One Cambridge resident saw a need for modern music to act as an 'electric shock', drawing the listener out from the 'slough of prejudice'.[37] 'These modern composers must enjoy something better than posthumous honour', the writer concluded.

The controversy was pursued by Scholes in an article, 'The Music of Today', which appeared in the *Radio Times* of 18 May 1928.[38] Modern composers were divided into two camps, he claimed: New Romantics and Anti-Romantics. Strauss, Elgar, Skriabin, and even Schoenberg were New Romantics. Stravinsky and Bartók were Anti-Romantics. As if to reassure his more conservative readers, Scholes meted out rougher treatment to the second group. In the section about Bartók he wrote: 'He is seeking to cast off the romantic clothing of the Nineteenth Century, and when in a broadcast programme he suddenly appears naked and unadorned, no wonder that some of us put our hands to our eyes and cry "Fie".'

This debate could only help Bartók's cause, appearing as it did in one of the most widely read national publications. As Bartók had commented several years before, a definite reaction, even a hostile one, was always preferable to polite inaction.[39] The BBC, too, was not to be deterred from its promotion of modern music by listeners' first reactions, and through Edward Clark willingly arranged two further concerts for Bartók, giving considerably more than the usual six weeks' notice.[40] This visit in early March 1929 was preceded by several performances of Bartók's latest quartets, which well prepared the public for other recent compositions which Bartók would perform with violinist Zoltán Székely. The BBC mounted a studio performance of the Third String Quartet (1927) on 12 February with the Vienna (Kolisch) Quartet, followed on 22 February by the world radio première of the Fourth String Quartet (1928), which preceded its Budapest concert première by nearly a month. The Hungarian (Waldbauer) Quartet, which gave these earliest performances of the Fourth Quartet, also performed the Third in the Wigmore Hall on 19 February.[41]

Before Bartók's arrival the BBC engaged in an exercise of public education, in an attempt to persuade listeners to accord a greater respect to Bartók's music and not to switch off their sets too unthinkingly. A short

---

[36] John J. Allen, 'It May Appeal to Savages, but—', ibid.

[37] B. St D. A., 'We Need Shocking!', ibid.

[38] pp. 285–6. Scholes left Britain for Switzerland later in 1928, after gaining a lucrative contract to provide informative notes for a pianola company. He only returned to live in Britain in 1940.

[39] Watson Lyle, 'Béla Bartók', MNH 19 May 1923, p. 496.

[40] BBcl pp. 442–3.

[41] See reviews: The Times, 20 Feb. 1929; MT 70 (1929), 352.

article appeared in the *Radio Times* for the week of the concerts entitled 'Béla Bartók: A Note on Monday's Distinguished Visitor'.[42] The writer pointed out that nothing could stand still without stagnating, not even music. Composers who were once perceived as radical and inspired a storm of abuse were now a part of 'the common man's delight'. The 'common man' then came to the punch-line:

It does behove us to be patient, and, moreover, to be humble. To our grandchildren, Bartók and the other 'fiery particles' of today may well be the kindly and inspiring friend that Wagner is to us; in *The Radio Times* of 1980, someone may be quoting Bartók as Wagner is cited here, by way of a sermon on the virtue of tolerance ... However little we may find to enjoy, or even to understand, in his unaccustomed idiom, we B.B.C. listeners can at least be honestly proud of his coming as an event in which we have a share.

On a later page, where the programmes of 4 March were listed, a central feature made quite explicit the listener's programme options for that evening: 'YOU TAKE YOUR CHOICE—between the music of Béla Bartók and Herman Finck in the programme tonight'.[43]

The programme on 4 March 1929 presented many of Bartók's most recent compositions—the two Violin Rhapsodies (both 1928), the Sonata (1926), three pieces from *Out of Doors* (1926), *Three Rondos on Folk Tunes* (1916/27)—along with some of his better-known earlier works. The concert was an 'outside broadcast', part of the BBC's third season of contemporary music held at the Arts Theatre Club in Great Newport Street. Only the first hour from 8 p.m. was broadcast. Bartók deliberately left his Sonata for piano until after 9 p.m., as he considered it 'too difficult for the radio audience'.[44] Alas, it was too difficult for Bartók as well! Howard Ferguson was in the audience and recalls:

[Bartók] later played the Piano Sonata in which, playing from memory, he came to a complete halt twice, his memory having failed. Each time he sat perfectly still for a moment, then started off again, apparently quite calmly (though I doubt he was as calm as he looked) at some point previous to that at which the disaster had occurred.[45]

Memorization of this work clearly caused Bartók—along with many pianists since—considerable difficulty. By 1933 he was preferring to play it with the score, and in 1934 is found trying to leave this work out of a programme on the grounds that it would give his audience a fright.[46]

---

[42] D. M. C., *RT* 1 Mar. 1929, p. 504.

[43] Ibid. 516. Finck was a British composer and conductor of light music. In contrast with Bartók, his most popular piece 'In the Shadows' had sold over a million copies.

[44] *BBlev*.v p. 358.

[45] Unpublished letter, Howard Ferguson to the author, 13 Apr. 1984.

[46] See respectively: Stuart Hibberd, *'This—is London'* (London, 1950), 91; *BBL* pp. 226–7.

Published reviews of this concert were remarkably uniform: general condemnation of Bartók's piano tone, but praise for the violinist Székely; surprise at the straightforward construction of the new rhapsodies, but difficulty in following the more recent piano pieces.[47] Two short sets of Hungarian folk-tunes and Romanian folk-dances arranged for violin and piano were found so light as to be almost drawing-room music. The most bitter words came, as often in the past, from the *Observer*:

We do not understand why it is a good thing to jump from one end of the keyboard to another into a resolution which does not resolve, or to cut off a quaver from a bar of common time and call it seven-eight. Such delights are too violent for us, too blood-and-thunder; life is not like that really—at least, not here. There is something parochial about it; the place for genius is in 'the stream of the world.'

Bartók's other BBC engagement was completed, without press comment, on the following evening, when he performed four of his transcriptions of Italian cembalo music in a late-night studio broadcast.

Before leaving Britain, Bartók was interviewed by Calvocoressi in the *Daily Telegraph*.[48] Encouraged by his friend, he abandoned his usual reserve towards the press and spoke openly on a wide variety of subjects. After giving an account of the controversial performance of his pantomime *The Miraculous Mandarin*, Op. 19 in Cologne in 1926, Bartók talked about his present tendencies in composition: his avoidance of sustained melodies and preference for short phrases within a polyphonic texture. He denied ever writing atonal music, but did confess sometimes to using certain devices of harmony or melody to veil the tonality. Later in the interview he attacked the widespread British hostility to Liszt's works, and declared his admiration for the standard of the orchestras with which he had played during his recent tour of the United States.

The following year, 1930, was the high point of Bartók's association with the BBC. During that year the BBC Symphony Orchestra was founded and Adrian Boult assumed the Directorship of the Music Department. Bartók was among the first to benefit from the bold, imaginative programming of orchestral works characteristic of the earliest years of Boult's tenure, when the blights of the Depression, cumbersome bureaucracy, and 'home preference' had not yet bitten too deeply into the organization. Others who would benefit from these few years of liberality included Stravinsky, Schoenberg, Berg, Prokofiev, and Strauss.[49] During 1930 Bartók came to Britain three times. His most important engagements were at the Queen's

[47] Reviews mentioned: E. B. [Eric Blom], 'Music by Béla Bartók', *MG* 5 Mar. 1929; Anon., 'Music of the Week', *Observer*, 10 Mar. 1929. See also: *DT* 5 Mar. 1929; *MT* 70 (1929), 351–2; *MMR* 59 (1929), 109.

[48] 'Music of the Day', *DT* 9 Mar. 1929. See also Malcolm Gillies, 'A Conversation with Bartók: 1929', *MT* 128 (1987), 555–9.

[49] See Adrian C. Boult, *My Own Trumpet* (London, 1973), 94–113.

Hall, where he took part in broadcast performances of his First Piano Concerto (1926) and *Rhapsody*, Op. 1, the latter inducing a new level of warmth in his reception from average concert-goers. Three further broadcasts were also given in chamber or solo roles. This high level of exposure was paralleled in the *Radio Times*, where Bartók was featured in short articles, often accompanied by photographs or caricatures.[50] Bartók's fees, too, reflected the importance officially ascribed to his visits. They totalled £189 for the five appearances.

Bartók's admission to conventional symphony concerts was negotiated by Sir Henry Wood. Although Wood had performed various of Bartók's orchestral works since 1914 he had never collaborated directly in performance with the composer. For the 1929–30 season of his concerts he conceived the idea of introducing Bartók's much-discussed First Piano Concerto to his Queen's Hall audience.[51] On 14 February the thirteenth of Wood's BBC Symphony Concerts took place, with Bartók's concerto commencing the second half, sandwiched between 'Weather and News' (for radio listeners) and a Tchaikovsky suite.[52] Despite his many visits to London by this time, Bartók had never before come face to face with its average music-goer. The majority of the Queen's Hall audience on 14 February, as well as the critics, had come to hear the well-loved Beethoven, Mozart, Brahms, and Tchaikovsky items in the programme. Would Wood's bright idea be accepted by this staid gathering?

Only a few people actually left the hall during the concerto's rendition, and polite if not voluminous applause greeted its end.[53] From the critics over the following days came numerous, mainly lacklustre, reviews. Many critics, being unversed in contemporary music, scrambled for indecisive clichés to paper over their feelings of bewilderment at the performance. They repeated, in some cases verbatim, general observations about Bartók's music which had been expressed in earlier days by supposed experts. As Basil Maine observed in the *Morning Post*:

There can be no doubt that this Concerto (written four years ago) achieves its end with great assurance. Critics are wont to say this of all Bartók music. Then comes the inevitable qualification—'Whether you like what has been achieved or not is another matter'. It is surely for the critic to make a decision on this point.

[50] *RT* 3 Jan. 1930, p. 24; 7 Feb., pp. 321 and 322–3; 14 Nov., p. 441; 21 Nov., pp. 518–19.

[51] Henry Wood Papers, BL Add. MS 56434, booklet 'Notes for Promenade Concerts—Season (35th) 1929'.   [52] *RT* 7 Feb. 1930, p. 349.

[53] Reviews mentioned: B. M., 'B.B.C. Symphony Concert', *MP* 15 Feb. 1930; 'London Concerts and Recitals', *MS* 22 Feb. 1930, p. 62; R. C., 'In the Concert Room: Bartók's Concerto', *MMR* 60 (1930), 79; H. F., 'The Week's Music', *ST* 16 Feb. 1930; Anon., 'B.B.C. Concerts', *MO* 63 (1929–30), 515; Anon., 'B.B.C. Concert', *The Times*, 15 Feb. 1930; M. S. J., *ES* 15 Feb. 1930; Anon. 'Bartók Béla sikere az angol rádióban', *Magyarság*, 16 Feb. 1930. See also: *Daily Chronicle*, 15 Feb. 1930; *MT* 71 (1930), 259.

Maine's decision was positive. In the *Musical Standard*, Watson Lyle also wrote favourably of Bartók's stimulating dissonance, technical resource in instrumentation, and logical treatment of thematic material. But they were countered by those few who had decided negatively. Richard Capell believed that Bartók 'in sacrificing to his barbaric Muse' had written a work rather more fierce than nature could possibly have intended. Critics from the *Sunday Times* and *Musical Opinion* questioned the very rationale for such a work. The latter suggested that it was 'a clashing cascade and cataclysm of chaotic cacophony'! For *The Times*'s writer the issue of generalist and specialist audience could not be ignored because of the nature of Calvocoressi's notes in the programme-booklet:

It is all very well for the well-instructed writer of programme notes to point to 'ingenious exploration of the possibilities of sonata form,' but people do not go to symphony concerts in the mood of hard endurance proper to an exploring expedition, and these enterprising composers will have sooner or later to go alone or at any rate with only such company as the 'Contemporaries' can offer them.

Between these two camps, as if paralysed, lay most other critics. The high-flown headline 'Greyhaired Ascetic as High Priest of Harmonic Percussion in Music' in the *Evening Standard* was a valiant attempt to disguise a neutral evaluation of the work as 'interesting'. In Hungary the concert was reported with the customary favourable headline, 'Béla Bartók's Success on English Radio', but in the smaller print this column told only of an 'exceptionally great interest' in the work from audience and press, and quoted solely from the *Morning Post*.

This fearful response to modern music from many of the London critics drew comment from the music publisher Hubert J. Foss in an article of the following April.[54] Observing a growing mass interest in music, nurtured by the gramophone and radio, he lamented the ineptitude of the BBC and the critics in harnessing that enthusiasm more effectively for the support of new music. Through 'manifest compromises' the BBC's greatest possibilities for musical good were largely unrealized, he maintained. Concerning the critics, Foss argued:

It has long been apparent . . . that the public is more receptive of new music than the critics are. Indeed, works have had popular success during the very period when official criticism was condemning their sanity. This is as true of the recent past as it is of the present; Bartók suffers precisely as Debussy did, and the extra-ordinary concentration on manner rather than matter still persists. . . . New music is still received with fear rather than with interest, with astonishment rather than perception, is condemned rather than expounded.[55]

[54] 'The Musical Press in England To-day', ML 11 (1930), 128–40.
[55] Ibid. 131.

Bartók and the BBC had not expected the concerto to gain a standing ovation in the nation's living-rooms. For a 'progressive' work, however, it had not scored badly in so conservative a milieu. Plans were drawn up for the performance of Bartók's only other work for piano and orchestra, his *Rhapsody*, Op. 1, which Wood had already conducted in 1921. The success of this work when it was finally produced on 26 November 1930 owed something both to thoughtful programming and to the performing excellence of the BBC Symphony Orchestra, which had given its first official performance only in the previous month. Bartók's composition was neither the most recent nor the most sensational in the programme, as it shared the second half with Ravel's *Boléro*.[56] Even before the concert Bartók realized that a good performance was in store, writing to his wife: 'The rehearsals are over—they were boring for me. The rhapsody is, of course, going well. With such a good orchestra it's little wonder (although things went even better in America).'[57] Wood's comments throw further light on the reasons for the quality of performance:

Bartók is certainly an original composer, but he is a trifle too fastidious for Promenade productions, for he demands more time than it is possible to give him at rehearsals. However, I have always made a rule of having piano rehearsals at my house a day or two before a concert containing works of these foreign composers. I look upon this as valuable, not alone in the musical sense but in the personal. By means of these rehearsals I have come to know these men *well*—in some cases intimately; and this so helps one's interpretation of a work.[58]

As with the reception of Bartók's *Two Pictures* in June 1928, the critics were taken aback by how pleasant Bartók's earlier compositions sounded. 'Rather genial and even mild', 'fresh', 'attractive', 'frank and tuneful' were expressions rarely heard in connection with Bartók's music, but now applied to his Opus 1.[59] Certainly the work was an early one, but at least ordinary British listeners were coming to appreciate compositions from his output which were more than simple folksong settings. Only the *Morning Post*'s critic expressed disappointment with the work, finding it too conventional and dull.[60] Bartók himself recognized the success of this rhapsody with mixed feelings, writing later to his mother: 'It is true that my success in London was rather great, but it came a little late—by approximately 24 years.'[61] Composed in 1904, submitted unsuccessfully

---

[56] MTA BH 2400/47. The first half of the concert consisted of works by Mozart, Strauss, and Tchaikovsky.

[57] Letter, Bartók to his wife, 26 Nov. 1930, *BBcl* pp. 497–8.

[58] *My Life of Music* (London, 1938), 427.

[59] See reviews, respectively: *DT* 27 Nov. 1930; *The Times*, 27 Nov. 1930; *ES* 27 Nov. 1930; *MS* 13 Dec. 1930, pp. 193–4.

[60] F. T., 'B.B.C. Symphony Concert', *MP* 27 Nov. 1930.

[61] Postcard, Bartók to his mother, 6 Dec. 1930, *BBcl* p. 500.

for the Paris Rubinstein Competition of 1905, and first publicly performed
in 1906, the work had waited until now for such recognition.

Although these Queen's Hall engagements were the most noticed, there
were other important BBC concerts for Bartók in 1930. On 6 January he
joined with fellow Hungarians József Szigeti and Mária Basilides in an all-
Bartók programme presented at the Arts Theatre Club. As in his last concert
at this venue, only the first half of the programme was broadcast. British
reviews were few, although Calvocoressi did bless the entire performance
with admiration, and the *Observer*'s critic, probably A. H. Fox Strangways,
commented briefly on Bartók's more direct expression in his latest works
and the generally high level of the performance.[62] A review appeared,
however, in Hungary, written by a correspondent from *Az Est* of Budapest
who had tuned in to the concert and, almost like a sports commentator,
graphically described its course:

Last night we listened to the Bartók concert in London transmitted through
Daventry. The first item was the new version of his violin rhapsody [the First]. It was
played by József Szigeti, to whom it was dedicated, with Bartók accompanying at the
piano. The audience at the Arts Theatre Club received the two artists with a loud
ovation. As soon as Mária Basilides appeared on the stage there was great applause,
which was repeated with renewed strength after each item. She sang four new
Hungarian folksongs of Bartók to their original Hungarian text. The excellent radio
reception made it possible to hear the words clearly. Some of Béla Bartók's own
solos then followed: the second Elegy and two Burlesques, performed in a masterly
fashion, then the second violin-piano sonata, in which Bartók's partner again was
Szigeti. This concert in London was a triumph for modern Hungarian music, and the
loud, long-lasting acclamation of the English audience proved that they understood
what the great Hungarian composer was saying, as interpreted by these three
brilliantly talented Hungarian artists.[63]

In the two remaining BBC studio engagements for 1930 Bartók presented
himself more as a pianist than as a composer. By the sustained promotion of
his own compositions in the past he had allowed his wider interpretative
powers to be overlooked. On 5 January he therefore presented a brief
programme of piano pieces by Purcell, Bach, Kodály, and Bartók, in an
early evening time-slot. Likewise, on the evening of 24 November he gave a
varied programme of items by Bartók, Mozart, and Kodály. For the central
item, Mozart's Violin Sonata in A major, K 305 he was joined by his former
associate in London, Jelly Arányi.

While in London for the BBC, Bartók found time to make his first English
recordings for the Columbia Gramophone Company. Ever since 1927
Szigeti had been trying to arrange a joint recording. In August 1929 he saw
the opportunity of bringing this about during Bartók's visit to London of

---

[62] 'London Concerts', *MT* 71 (1930), 167; Anon., 'Concerts of the Week', *Observer*, 12
Jan. 1930.   [63] *Az Est*, 8 Jan. 1930.

January 1930.[64] It was important, Szigeti realized, to deal with companies in London rather than through their European representatives, as only this direct approach would ensure the placement of a record in catalogues throughout the world.[65] On 7 January the Bartók–Szigeti *Hungarian Folk Tunes* and Bartók–Székely *Romanian Folk Dances* were recorded by the duo at the Abbey Road studios of Columbia.[66] These works, likened to 'modern drawing-room pieces' during the previous year, were well chosen for the market. When released later in 1930 they gained good reviews. Of the Romanian pieces the *Gramophone* commented: 'These will be liked for their odd flavour and neat build. They are quite easy to listen to, and the record is extremely clearly made: quite one of the best of the month.'[67] Szigeti nursed Bartók through the entire process, even making sure that he wrote to gain his free records when they were released.[68] Thirty years later Szigeti would look back on these London days and lament that through Bartók's inability to promote himself all too few professional recordings had been made of his playing.[69]

Bartók did not visit Britain in 1931 but came in the subsequent years, each time for a British première with the BBC of one of his major orchestral works: *The Miraculous Mandarin*, Op. 19 (concert version) in 1932, the Second Piano Concerto (1931) in 1933, *Cantata Profana* (1930) in 1934. Through Edward Clark's adventurous programmes, backed up by propaganda in the *Radio Times*, the *Listener*, and broadcast commentaries, the BBC sought slowly to raise the level of appreciation of Bartók's works so painstakingly established on his earlier visits. This use of a national radio service for the furtherance of modern music was envied by many across the Channel and the Atlantic. In 1934 Eyvind H. Bull, the father of one of Bartók's piano students at the time, remarked in the *Music News* of Chicago:

We know amazingly little about Béla Bartók ... We Americans have not been so fortunate as the 'unmusical' English; for in England the British Broadcasting Company [sic] first performed his piano concerto about seven years ago (I believe with the composer at the piano) and with real courage and good sense has continued to perform his works until now there is a large public truly interested in Bartók's compositions.[70]

When *The Miraculous Mandarin* suite was introduced to British listeners on 4 March 1932 it was cunningly placed after two early Bartók works of

[64] MTA BH 1563.   [65] MTA BH 1566.

[66] Col. LX31 and Col. LB6. The recordings are included in *Bartók At The Piano* (Hungaroton, LPX 12328–B).

[67] W. R. Anderson, 'Analytical Notes and First Reviews', *Gramophone*, 8 (1930–1), 277.

[68] MTA BH 1569.

[69] 'Working with Bartók', *Music and Musicians*, 11. 8 (Apr. 1963), 8. Szigeti tells more broadly of his years of collaboration with Bartók in *With Strings Attached* (New York, 1947).

[70] 'Pointed Paragraphs', *Music News*, 7 June 1934, pp. 9–10.

proven popularity, the *Suite* No. 1, Op. 3 and the *Rhapsody*, Op. 1, the latter featuring Bartók as soloist. Because of shortage of space the performance was held in Studio 10, a disused wine warehouse near the southern end of Waterloo Bridge. Bartók later described the venue as a 'kind of hangar'.[71] The BBC had prepared well for the concert. Frank Whitaker's racy article 'The Most Original Mind in Modern Music' was included in the week's *Radio Times*.[72] With journalistic flair Whitaker told of Bartók's incurable nervousness when on stage (including his one abortive engagement as a conductor), his vast technical knowledge of music, and his steady determination to write what his conscience rather than his public dictated. Midway through the article, however, its educative aim became apparent, albeit in a more subtle fashion than in earlier BBC writings: 'Progress always means more complications. There are traffic signals in Piccadilly Circus, and on the sideboard plain gin has given way to the cocktail. The physical ear of a Bartók or Stravinsky is simply a higher organism than yours or mine. What is tobacco to us is bread and butter to them.'

Reviews of this concert on 4 March 1932 reveal that while Bartók's early works were able to gain a modest measure of acceptance from the public and generalist critics, they were increasingly under attack from the specialist critics of modern music. To Herbert Hughes the *Rhapsody*, Op. 1 was now 'naive and childlike'; to Richard Capell it was just a non-starter, whereas for Edwin Evans its very direction was hard to follow.[73] This disparagement of the early *Rhapsody* did not mean, however, that his more adventurous and later *Mandarin* was unconditionally accepted. While Hughes found the suite 'vastly entertaining', Capell wrote of its 'shocking din', and Evans questioned its real value:

Taken seriously it might border on the unpleasant. Stripped of pretensions it is sheer melodrama, and since Studio 10 is on the Surrey side one was prepared to meet it on that basis. For this mine-drama Bartók has composed appropriate blood-and-thunder music. . . . It makes Salome in retrospect nothing worse than a saucy minx. Such music can be, as this is, thrilling to hear once in a while. Its permanent value, detached from its subject, is more problematic.

None the less, when writing a 'Homage to Sir Henry Wood' in 1944, Bartók recalled this Studio 10 concert with affection: 'This was a "landmark" in my career: it was the first orchestra concert devoted entirely to my works.'[74]

In 1933 it became clear that the BBC's campaign for greater acceptance of

[71] *BBE* p. 521.   [72] 26 Feb. 1932, p. 504.
[73] Reviews mentioned: H. H., 'Béla Bartók's Music', *DT* 5 Mar. 1932; R. C., 'Hungarian Concert', *DM* 5 Mar. 1932; E. E., 'Contemporary Music', *MT* 73 (1932), 359.
[74] *BBE* p. 521 (edited text). Bartók is in error. The BBC broadcast on 10 October 1927, for instance, had been devoted solely to three of his works. During this 1932 visit, on 2 March, Bartók also performed in a little-publicized evening broadcast. He played works by Bach, Scarlatti, and Marcello.

Bartók's music had, in essence, stalled. Although there had been a repeat performance of the *Mandarin* suite on 26 February, his music was being played less than in the heady days of 1929 and 1930.[75] A limited number of professional musicians were certainly coming to appreciate the full measure of his talents, but average music-lovers were voting with their feet or radio dials and snubbing the performances of his most recent compositions.[76] When the composer played his Second Piano Concerto (1931) on 8 November at the Queen's Hall, the more informed of the critics were full of praise. But the audience reaction was icy: 'They might at least have had the energy to hiss,' commented the *Monthly Musical Record*, 'but not even this token of interest was forthcoming. Instead, nothing but weak and lukewarm applause—isolated hand-claps in a desert of approval.'[77] In *The Times* a correspondent conjectured that the inclusion of Bartók's concerto was probably responsible for the poor attendance at the concert, notwithstanding the more conservative works by Rimsky-Korsakov and Bliss which were also on the programme.[78]

For the reviewers the most striking feature of this new piano concerto was its unexpected Romanticism. Most took exception to the programme-notes, written by Edmund Rubbra, which had, following the lead of Percy Scholes, proclaimed Bartók as the 'supreme anti-romantic composer'.[79] In this new work it was felt that the grim, uncompromising side of Bartók's art was less evident. Ernest Newman was surprised to hear occasional melodic formulas straight out of the Romantic era, and commented on the different directions being pursued by the melody, harmony, and rhythm in the work.[80] While Bartók's violin sonata of 1921 had been castigated by Newman for its ugliness and incoherence, his piano concerto of 1931 was found to be interesting and enjoyable. In short, its 'bark is considerably worse than its bite'. Newman pushed the dog analogy even further:

For a moment we may get a jump of the heart and an horripilation of the skin as something that appears to be a lion leaps out at us from behind the hedge; but when we see that it is only old Rover, the leonesquely-barbered poodle from Honeysuckle

---

[75] See Hubert J. Foss, *Music in My Time* (London, 1933), 151–2. There was also less consideration of Bartók in the British press during the early 1930s than in the preceding years. 1931 was something of an exception, however, because of Bartók's fiftieth birthday and the appearance of his volume *Hungarian Folk Music*. See: László Pollatsek, 'Béla Bartók and His Work', *MT* 72 (1931), 411–13, 506–10, 600–2, 697–9; Arthur G. Browne, 'Béla Bartók', *ML* 12 (1931), 35–45; Bernard van Dieren, 'Musical Microtomy', *MMR* 61 (1931), 300–3.

[76] Calvocoressi discussed this situation in 'London Concerts', *MT* 74 (1933), 1128.

[77] Anon., 'Notes of the Day', *MMR* 63 (1933), 221. In a letter to the author of 13 Apr. 1984 Howard Ferguson recalled this reception as one of 'politeness rather than wild enthusiasm'.

[78] Anon., 'Béla Bartók at Queen's Hall', *The Times*, 9 Nov. 1933.

[79] Programme booklet, pp. 6–12, MTA BH 2400/91. The notes were also published as 'Béla Bartók's Second Piano Concerto', *MMR* 63 (1933), 199–200.

[80] Reviews mentioned: 'The Week's Music', *ST* 12 Nov. 1933; R. C., 'In the Concert Room', *MMR* 63 (1933), 227; M.-D. C., 'London Concerts', *MT* 74 (1933), 1128.

Cottage down the lane, we merely pat the friendly and likeable beast on the head and tell him we look forward to our next merry meeting with him—as we do with this 'grim and uncompromising' concerto of Bartók's.

Capell also had dogs in mind. Having examined Bartók's handling of musical elements he voiced the suspicion 'that Bartók cannot be quite so sad, mad, bad a dog as he makes himself out'. Bartók's ferocity had been curbed a little, Capell thought, and so he looked forward to rehearing the work so that he could verify if this was really the case or whether it was just an illusion created by Adrian Boult's skilful conducting. As was now customary among the critics, it was left to Calvocoressi to utter the kindest words. To him Bartók's latest piano concerto was 'pure joy from beginning to end'. Its cool public reception was not, however, and Calvocoressi saw a remedy to this only in a number of first-rate pianists including the work in their concerts. Although at first they would receive little thanks, he maintained that eventually the composition would be propelled into the front rank of the concerto repertory.

Bartók's low popularity was not for want of trying on the BBC's part. In the *Radio Times* of 3 November he had been featured with a photograph and a small article.[81] Yet the article's conclusion would have done little to allay the fears of the middle-of-the-road listener: 'His art is a law unto itself, and Bartók differs from most of his contemporaries in being a slave to nothing but his deliberate and reasoned intentions. He has for his own purposes deposed law and set up logic in its place.' The following day in a relay transmission from Budapest the BBC broadcast the *Rhapsody*, Op. 1, with Bartók's pupil Lajos Hernádi as soloist. This was a prelude to a London studio appearance by Bartók himself on the evening of 6 November. Broadcast at 9 o'clock over London, North, and Scottish regional stations, the programme consisted of Bach's Suite in G minor, BWV 808 and Bartók's Sonata (1926) for piano. Given his reticence, both earlier and later, about performing this sonata, its inclusion is surprising. Advice from the BBC did little to foster a relaxed performance: 'As this programme looks as if it might run a little too long, would you please do anything in your power to hurry it up?'[82] Stuart Hibberd, a leading BBC announcer, recalled the performance in his memoir '*This—is London*':

[Bartók] was a tall [*sic*], thin, clean-shaven man in the middle fifties, with a rather lined face and a pleasant smile. He began by playing some Bach from memory, then he played a piano sonata of his own, in three movements, and for this he asked me to turn over. This was not too simple a matter, because the music was in manuscript and not very clear, and he proceeded to use the piano as a percussion instrument, and in some of the louder passages lifted his hands so high that it was difficult to get

[81] Anon., 'Béla Bartók', *RT* 3 Nov. 1933, p. 343.
[82] Unpublished letter, KAW [Wright] (BBC) to Bartók, 27 Sept. 1933, BBCWAC 47796.

near enough to the music to follow it. It was all very exciting, almost thrilling at times, but I was thankful when it was all over and I was able to relax once more.[83]

On the other end of the transmission, 'Audax', a wireless correspondent for the *Musical Times*, was listening, and was none too impressed.[84] The sonata sounded to him like a 'lot of dwarfs brewing potions', and the Bach suite was criticized for the uneven rhythm in one of its movements. In a parting shot this critic questioned the wisdom of importing such foreign artists as Bartók.

The final work in the BBC's trilogy of Bartók novelties was his *Cantata Profana: The Nine Enchanted Stags* (1930). Transmitted on 25 May 1934 from the Concert Hall of the new Broadcasting House, this was a world première performance, although that fact escaped mention in promotional material. Preceding the cantata were the *Two Portraits*, Op. 5 and the Second Piano Concerto (1931), with Bartók as soloist. Some spare time at the end of the broadcast was filled in by Bartók playing a number of Hungarian peasant dances.[85] Because of the difficulty of the works a generous allowance of rehearsal-time was afforded over the preceding days, culminating in nine hours of full orchestral rehearsals on 24 and 25 May.[86] The performance also called generously on personnel: seventy-one players of the BBC Symphony Orchestra, forty-two members of the Wireless Chorus, as well as soloists and the conductor, Aylmer Buesst. Bartók was pleased with these arrangements and, overall, with the standard of the performers. He wrote home on the day before the concert:

We kept on rehearsing yesterday and today. The orchestra and choir are excellent. The baritone is also very good; the tenor less so. But the conductor is, for sure, nothing wonderful; he's really only a time-beater. He drags out the tempos. But even so, owing to the excellence of the others, one can gain some kind of picture of the cantata. It's most fortunate that I can listen to it for the very first time with such a good orchestra and choir.[87]

The response to the work was disappointing, despite its extensive advertisement[88]—the BBC had even commissioned a new cubist caricature of Bartók for the event. Whether because of lack of interest or the distraction of Gustav Holst's death on that day, the world première of Bartók's work gained little notice in the press, and most of that notice was unfavourable. *The Times*'s critic found his 'angular idiom' suitable in the

[83] (London, 1950), 91–2.
[84] 'Wireless Notes', MT 74 (1933), 1084–5.
[85] BBCWAC 47796 (25 May 1934).
[86] BBCWAC 47796 (16 May 1934).
[87] Letter, Bartók to his wife, 24 May 1934, *BBcl* p. 542. The baritone was Frank Phillips; the tenor was Trefor Jones.
[88] See: RT 18 May 1934, pp. 518, 567; *Listener*, 23 May 1934, p. 883; *The Times, DT*, and MG 25 May 1934.

BELA BARTOK

as the caricaturist sees him.
The B.B.C. Contemporary Music
Concert at 21.00 will be de-
voted to his music. Bartok will
himself play his Pianoforte
Concerto No. 2.

Full details of the concert will be
found below, and an article on the
Cantata Profana on page 518.

'Béla Bartók' (Radio Times, 18 May 1934)

concerto, but not appropriate to vocal music.[89] In the *Daily Telegraph*, too, Richard Capell expressed the opinion that the concerto (on a second hearing) had been considerably more effective than the cantata. Some of his reasons related to the cantata's performance—for technical reasons the singers' backs had been turned to the audience; the words had not been distributed—but others to the subject of the work, which Capell found 'bizarre and rather pointless'. Bartók's music was not directly considered. As ever, it was Calvocoressi who wrote in detail about the musical qualities of the work, finding greatness where others found only anguish.

Bartók's visits to London for these BBC engagements during the early 1930s were always brief, leaving little time beyond that required for rehearsals, performances, and occasional negotiations with Oxford University Press. For accommodation he usually accepted the standing invitation issued by the Wilsons of South Kensington, with whom he had first stayed in 1922, and only stayed elsewhere if their guest-room was already occupied.[90] From time to time he did manage to socialize with old friends, such as Adila Fachiri, Edward Clark, and Cecil Gray,[91] but these

[89] Reviews mentioned: Anon., 'B.B.C. Concert', *The Times*, 26 May 1934; R. C., 'New Bartók Works', *DT* 26 May 1934; M.-D. Calvocoressi, 'Lettre de Londres', *Revue musicale*, 15 (1934), 300.
[90] Bartók's relations with the Wilsons are discussed in Chapter 6, below.
[91] See, respectively: *BBcl* pp. 497–8; *BBcl* pp. 525–6; MTA BA-B 3477/67 and MTA BH 541.

meetings had to be sandwiched between the demands of his tight schedule. Only in March 1932 did he manage to gain a completely free day in London, and he revelled in the chance to be a tourist: 'I have at last "seen" something in London (the last time this happened was ten years ago) . . .'.[92] What did Bartók do? We know that he took tea and a plain bun at an ABC teashop, paid threepence for the pleasure, and thought this worth recording, along with the fact that he did not have to give a tip.[93] In preparation for his visit to Cairo later in the month for a Congress of Oriental Music he bought some Greek and Egyptian currencies, before spending time at an exhibition of French painting, which gathered together material from all over the world.[94] He found the display too large to view thoroughly, and was irked by the crowds which made movement difficult. So he focused on late nineteenth-century works, being particularly impressed by some Cézanne and Manet pictures.[95]

As the 1930s progressed Bartók enjoyed his visits to Britain more, but his connections with the BBC less. Broadcasting was becoming more professional; the BBC was becoming more bureaucratic, with greater specialization in job function and consequent loss of personal treatment. There was increasing discomfort in the relationship between the younger, practical staff of the BBC's Music Department and the powerful Music Advisory Committee, which was chaired by Sir John Reith and included such leading music academics as Sir Hugh Allen and the Master of the King's Musick, Sir Walford Davies.[96] Among performing musicians competition for radio exposure grew ever more intense and this led to factionalization within the Music Department itself, largely over the issue of artistic protection. By the mid-1930s the pressure for a 'British artists first' policy was becoming difficult to withstand.[97] The danger of such a policy had, none the less, been spelled out in the BBC's own journal the *Listener*: 'Music is an international art, and while it is right that we should be the first to recognise and support our own musicians, we would only be doing them a disservice by any attempt to exclude their foreign "competitors". Our first consideration must be to attract the best in music, from whatever source.'[98] In reality, it was becoming irritatingly difficult for Bartók and most Continental musicians to gain BBC contracts. Even in possession of these contracts, it

[92] Letter, Bartók to his wife, 2/3 Mar. 1932, *BBcl* pp. 525–6.
[93] *BBcl* p. 526.   [94] *BBcl* pp. 525–6.
[95] For details of Bartók's wider artistic interests see Béla Bartók, jun., 'Bartók and the Visual Arts', *NHQ* no. 81 (Spring 1981), 44–9; János Breuer, 'Bartók and the Arts', *NHQ* no. 60 (Winter 1975), 117–24.
[96] See Boult, *My Own Trumpet*, pp. 94–5; Lutyens, *A Goldfish Bowl*, pp. 126–35.
[97] See, e.g., letter, Boult to Alban Berg, 27 Feb. 1935, reproduced in Nicholas Chadwick, 'Alban Berg and the BBC', *British Library Journal* 11 (1985), 54: 'As you know, there is always a certain pressure on the B.B.C. in favour of engaging English singers . . .'.
[98] Anon., 'The Listener's Music: The Musical Capital of Europe', *Listener*, 2 Oct. 1935, p. 591.

was a battle to maintain reasonable levels of fee. Despite Boult's conciliatory talents, the conflicts of opinion at this time concerning the direction of BBC music were loudly voiced before the Ullsworth Committee in 1935. Reith was dismayed at the bitterness, narrow-mindedness, and even ignorance of many making representations, and found it necessary to devote many pages of his memorandum of rejoinder to 'the discords of musicians'.[99]

A dampening of enthusiasm in the BBC for mounting Bartók performances can be traced from the early 1930s. In 1932, after making lavish promises of concert and studio performances, the organization issued a formal contract for only one studio appearance. Bitterly Bartók wrote to his wife towards the end of January 1932:

In August they promised me all manner of wonderful things: an orchestral concert with the First Suite, Mandarin; a piano concerto, as well as a studio appearance. The result of all that: a performance of the rhapsody in a studio orchestral concert. They don't do anything about the other promised studio appearance because 'surely I must come to England anyway to play in Glasgow'. Though I only agreed to the Glasgow business for a ridiculous fee because the radio promised two appearances. In short—I will not play any more on London radio.—I would very much have liked to call off the whole thing right now, but it would have caused too much trouble. Therefore it's still on (on 4 March), but then, that's it!! This concertizing profession becomes ever more difficult.[100]

After considerable hassle some of the original plan was resuscitated and a second appearance was arranged. The last-minute arrival of his labour permit further annoyed Bartók,[101] but he could not afford to implement his rash threat and turn his back on the BBC. A sign of something more than inefficiency became evident in the planning of Bartók's 1933 tour. The BBC again only offered one engagement, and this time was clearly trying to break the established pattern of paired appearances.[102] After a voluminous correspondence, followed by much internal note-swapping within the BBC, a second engagement was offered—but against the better judgement of many in the organization, who were aware of the inroads being made into BBC audiences and players by Beecham's popular new London Philharmonic Orchestra.[103] The mail-bags had been telling the BBC for some time that the public did not appreciate many of the broadcasts of Bartók's music. Concert attendances told a similar story. The organization was seen in the press to be irrationally faithful to Bartók in the face of such manifest unpopularity.[104]

---

[99] See J. C. W. Reith, *Into the Wind* (London, 1949), 229.
[100] Letter, Bartók to his wife, 30 Jan. 1932, *BBcl* p. 520. Bartók's Scottish concerts are examined in Chapter 6, below.
[101] *BBcl* p. 526.   [102] BBCWAC 47796 (15 Aug. 1933).
[103] See BBCWAC 47796 (2 Aug.–27 Sept. 1933); Nicholas Kenyon, *The BBC Symphony Orchestra* (London, 1981), 86–7.
[104] Anon., 'From London', *Evening Citizen* (Glasgow), 3 Nov. 1933.

'Audax' in the *Musical Times* spoke for many when he cited the promotion of Bartók's music as a typical example of the misguided direction of the BBC's music policy: 'The B.B.C. is painfully bent on duty. Sometimes it reminds me of "I will have smiling faces around me, if I have to thrash every one of you to get them." '[105] In 1934, again, only one engagement was offered, for the usual fee of 40 guineas. This time no second concert was arranged. The implication was inescapable. Bartók became depressed at the downturn in his fortunes, writing in September of that year that he could not see even the slightest prospect of further performances in Britain.[106] He did not visit the country in 1935.

Bartók's engagements with the BBC in 1936–8 were only realized through the tenacity and cunning of his new representative for England and Holland, Antonia Maier (later Kossar). Sometimes using half-truths—occasionally plain lies—she battled valiantly with the BBC administration to secure engagements and to gain commitments beyond the usual two-month planning schedule. It was evident to Bartók and his agent that the BBC was not tightening up on performances of his music by others so much as on performances *by him*. A series of broadcasts of his music had taken place during the last months of 1935,[107] but for Bartók himself, after three months of negotiations, the BBC could only offer one London appearance: a repetition of his Second Piano Concerto (1931) on 7 January 1936. A ten-minute bracket of solos was also requested, but with the advice: 'maximum ten minutes, popular and attractive type, not Liszt, but preferably own music'.[108] A 40 guinea fee was offered, but Maier asked for 50, arguing that Bartók was having to come to Britain for only one concert.[109] In an internal BBC memo K. A. Wright, the Assistant Director of Music, suggested that the difference be split, and stated: 'It is also certain that we do not want to offer him a second date during this visit, either alone or with Székely.'[110] Maier had already foreseen a way around such a position. She suggested in a letter of 26 October 1935 that Bartók might give a broadcast over Midland Regional Radio, Birmingham, about the time of an engagement which he had accepted with the Liverpool Music Society.[111] It was a perceptive move.

---

[105] 'Wireless Notes', *MT* 74 (1933), 1085.　　　　　　[106] *BBlev*.v p. 485.

[107] At the 1935 Promenade series Bartók's *Hungarian Peasant Songs* (1914–18) had been performed in their orchestral version. There were other broadcasts of Bartók's works on 27 September (a relay broadcast from Dohnányi's Budapest home—see Plate 11), 10 October (Second Violin Sonata), and 1 December (Fifth String Quartet).

[108] Notes on unpublished letter, Maier to BBC, 6 Nov. 1935, BBCWAC 47796. The advice against playing Liszt was probably given because of the items by Liszt already in the evening's programme.

[109] BBCWAC 47796 (26 Oct. 1935).

[110] Unpublished BBC memorandum, Wright to Booking Manager (Music), 1 Nov. 1935, BBCWAC 47796.

[111] BBCWAC 47796 (26 Oct. 1935). Details of the Liverpool concert on 16 January are given in Chapter 6, below.

While London was deluged constantly with concert offers from famous artists, the other cities were not. The Midland Regional Director was excited by the idea, making a prompt offer of one radio concert, later increased to two, involving an additional fee of 60 guineas.[112] For both concerts, on 17 and 18 January, Bartók was engaged to play his Second Piano Concerto, with the BBC Midland Orchestra conducted by Leslie Heward.

After some years' absence, Bartók's return to the Queen's Hall on 7 January 1936 was propitious. The performance of his piano concerto was well received by the audience and generously reviewed in the press.[113] The programme-notes written by D. Millar Craig probably helped, as they pointed out in simple terms how Bartók's composition was formally based on classical models.[114] Readers of the *Evening Standard* had also been treated to a sympathetic sketch of Bartók's character a few days before the concert.[115] That article concluded:

Bartók's uncompromising ideals and fiercely individual style have delayed popular recognition. Neither is his personality ingratiating. He is unimpressive in appearance, of an inconspicuous presence and embarrassingly shy. When visitors address him as 'Master' he invariably replies: 'My name is Bartók.'

By contrast with London, the Birmingham studio performances of this concerto attracted no extended press comment.

In accepting the engagement in London on 7 January Bartók had cut his schedule rather finely, since he was due to play in Utrecht on the following day. The boat would not arrive in time, so Bartók took to the skies, for the only time in his life, aboard a KLM Fokker XXII. During the flight he wrote to his mother:

I am writing this card aboard this aeroplane (overleaf) at a height of 2,400 metres, in glorious sunshine, above an endless field of fleecy-frizzy clouds. So I have tried this as well! But it was worthwhile. The take-off was a little frightening to a beginner like me, but I now feel everything is peaceful and secure! Many kisses, Béla.[116]

A few days later, when about to return to Britain—by boat—he looked back on this adventure: 'The flight was wonderful. I should like to go everywhere by air from now on, but I shall only do so if it's a necessity.'[117]

In early 1936 Bartók's relations with the BBC appeared to be becoming

[112] BBCWAC 47796 (8 Nov., 2 Dec. 1935).
[113] See reviews: *ES* 8 Jan. 1936; *ST* 12 Jan. 1936; *DT* 8 Jan. 1936; *Pesti Napló* (Budapest), 9 Jan. 1936; *Esti Kurír* (Budapest), 9 Jan. 1936. None of these reviews threw significantly new light on the work.
[114] Programme-booklet, MTA BH 2400/130. Craig's analysis was largely drawn from Rubbra's earlier, more complex text.
[115] Stephen Williams, 'Béla Bartók is Coming Here to Play', *ES* 3 Jan. 1936.
[116] Postcard, Bartók to his mother, n.d. [8 Jan. 1936], *BBcl* p. 549.
[117] Letter, Bartók to Etelka Freund, 13 Jan. 1936, *BBlev.v* p. 519.

more healthy. Its Director of Music, Adrian Boult, on 15 March gave the British première of his Four Orchestral Pieces, Op. 12, a performance promised by Wood as far back as 1922. The following week *Cantata Profana* (1930) was repeated, now broadcast from the Queen's Hall with Beethoven's 'Choral' Symphony as its companion. In April the BBC Symphony Orchestra visited Hungary and performed in Budapest's Városi Színház. Bartók's Four Orchestral Pieces comprised the second item on the programme, as a sign of British respect for his talent. In an interview conducted by László Ráskay of *Pesti Napló*, Boult spoke of 'the popularity of Bartók and Kodály in England. In his opinion every public is basically conservative, but curiously, the English, who hardly respond to new tendencies in other branches of the arts, understand, enjoy and love modern music.'[118] Boult's sentences of international affability did not, however, mention the drama behind the programming of Bartók's work at this concert, which had led to the resignation of Edward Clark. Because of a policy of including a British work in every concert of the orchestra's European tour, as well as certain items requested by local concert officials, it looked as if there would be no room for a piece by Bartók, whose compositions, Clark realized, were largely ignored at this time in Hungary. As Clark's wife, Elisabeth Lutyens, has recorded: 'Rather than betray his old friend Bartók, Edward had, no doubt precipitously and in a rage, tendered his resignation.'[119] The Four Orchestral Pieces were reinstated, but Clark was not. Bartók had lost his chief supporter inside the BBC.

At the top the BBC may have been receptive to Bartók's music and performing needs—Boult had even telegrammed Bartók for some programme-advice in November of the previous year[120]—but lower down a harder line was being forged. On 21 July 1936 Bartók's agent, Antonia Kossar (née Maier), wrote to the organization giving notice that Bartók was free to play in England towards the end of the coming January.[121] She continued: 'Mr. Bartók offers you the first performance in the world of a small series of new piano-pieces he has composed. They last together about 30 minutes. Would that interest you?' Keeping in mind the recent engagements in Birmingham, she suggested that further BBC appearances might be arranged in other cities, such as Glasgow and Belfast. Julian Herbage, an Assistant in the Music Department, snapped back in an internal memo:

This is surely another instance of artists trying to make their visits to this country dependent on what we can promise them in the way of engagement. We feel that we could only consider booking Bartók if he were already over here, and then it would

[118] 'Interjú Adrian Boult karmesterrel a londoni próbák idejéből', *Pesti Napló*, 18 Apr. 1936.   [119] *A Goldfish Bowl*, p. 135.   [120] MTA BH 213.
[121] Unpublished letter, Kossar to BBC (Wynn), 21 July 1936, BBCWAC 47796.

probably be for his solo pianoforte pieces, plus possibly a recital with Waldbauer if he is also over.[122]

With some rhetorical sweetening the message was conveyed to Kossar: the BBC was no longer prepared to be an initiator of a visit, but might be willing to contribute an engagement to a tour which was already firmly planned.[123] Bartók lamented the state of affairs with the BBC, in a letter to Kossar expressing his frustration that all dealings with that organization were so difficult and habitually left to the last minute.[124] But Mrs Kossar was more practical. She simply asserted to the BBC that Bartók and Székely did have engagements in Manchester on 3 and 9–10 February.[125] The BBC fell for what appears to have been a ruse, and an engagement was made for the duo to give a late-night broadcast of Bach and Bartók works on 4 February for a shared fee of 60 guineas.[126] Apparently after the confirmation of this performance Bartók arranged a further engagement, on 9 February, before the British section of the International Society for Contemporary Music, at which the various *Mikrokosmos* pieces originally offered to the BBC were given a world première.[127] While the BBC had shown no interest in this new collection of piano pieces, the Columbia Gramophone Company did. It contracted immediately with Bartók to record two numbers, 'Staccato' (No. 124) and 'Ostinato' (No. 146) at its studios on 7 February.[128] There were no concerts in Manchester!

Relations with the BBC did not improve in 1937. At a Programme Committee meeting on 1 April a proposal from the Music Department for a recital of pieces from Bartók's *Forty-four Duos* (1931) for unaccompanied violins was rejected. Mention was made of their 'very limited appeal'.[129] In August an offer from Bartók and Székely to give the world première of the (Second) Violin Concerto was graciously declined. K. A. Wright explained to Bartók in a personal letter: 'Unfortunately, when the final reviewing of the programmes was undertaken and the plans of old and new, familiar and unfamiliar re-made, it was not found possible to retain your new Concerto in the scheme.'[130] As matters turned out, a positive reply would have embarrassed Bartók. He had only started composing the concerto, and was not to finish it until the end of 1938. For Bartók's *Music for Strings, Percussion, and Celesta* (1936), which had already been premièred most successfully in Basle, Wright was none the less keen to secure the first British

---

[122] Unpublished BBC memorandum, Herbage to Minns, 25 Aug. 1936, BBCWAC 47796.
[123] BBCWAC 47796 (27 Aug. 1936).   [124] KBKK p. 169.
[125] BBCWAC 47796 (13 Nov., 26 Nov. 1936, 12 Jan. 1937).
[126] BBCWAC 47796 (11 Dec. 1936).
[127] This concert of 9 February 1937 is discussed in Chapter 6, below.
[128] Col. DB1790. The recording is included in *Bartók At the Piano* (Hungaroton, LPX 12326–B). See also *BBcl* pp. 572–3.
[129] BBCWAC 47796 (6 Apr., 12 Apr. 1937).
[130] Unpublished letter, BBC (Wright) to Bartók, 19 Aug. 1937, BBCWAC 47796.

performance, although he could only offer a time in the 1938–9 season. (The work was actually performed earlier than this, at a BBC Contemporary Concert on 7 January 1938, with Hermann Scherchen conducting.)

Having passed up the apparent first performance of Bartók's *Mikrokosmos*, the BBC was surprised to be offered by Kossar 'the first complete performance' of the work, not having realized its large dimensions. It was interested in such a performance, an official declared, but only under usual BBC conditions.[131] On 4 October 1937 Kossar wrote asking for a commitment to an exact date in January, 'as Mr. Bartók has work in England during this time and therefore has to come anyway'.[132] The detailed correspondence between Kossar and Bartók at this time reveals no engagements.[133] But the lie did not produce results as immediately this time, since the BBC's Booking Department insisted that it would only confirm a date two months beforehand.[134] A few days later, however, an engagement was offered for 20 January.[135] This was to be Bartók's last formal BBC performance: a single mid-evening recital on the National network presenting three short brackets of *Mikrokosmos* pieces, interspersed with two selections from his previously rejected violin *Duos* played by Frederick Grinke and David Martin. From the professional critics the broadcast elicited no substantial reaction, although the responses of some listeners have been preserved, including this one from a Surrey resident:

Re Béla Bartók tonight. What absolute nonsense. Broadcast it to the Arabs by all means, but for God's sake let us have at least sanity in the programmes. . . . Gramophone celebrity records please & let Béla Bartók join Béla the communist, so go back to Hungary & stop there. We don't want education, we want diversion.[136]

Although Bartók had performed officially for the last time with the BBC, a considerable amount of negotiation continued to be carried out by his agents with the organization. It is a sad tale of bickering, delay, and disappointment, of continuing persistence on Bartók's side, and bluntly expressed intransigence from the BBC. The first issue to arise concerned the ISCM Festival held in London from 17 to 24 June 1938. The BBC decided that Bartók could not be engaged for any extra broadcasting work when in London to take part in the Festival, for this might be seen to discriminate against other visiting ISCM members.[137] Again, when approached by the Festival organizers for half the fee of £50 to be given to Bartók and his wife for the British première of the Sonata for Two Pianos and Percussion (1937), the BBC hesitated, although once it had decided to broadcast the

[131] BBCWAC 47796 (17 Sept. 1937). In fact, Bartók continued to add to the collection until 1939, by which time there were 153 pieces.
[132] Unpublished letter, Kossar to BBC (Wynn), 4 Oct. 1937, BBCWAC 47796.
[133] See KBKK pp. 177–83.   [134] BBCWAC 47796 (25 Oct. 1937).
[135] See KBKK p. 184.   [136] Letter, Ew. R. to BBC, 20 Jan. 1938, *DB*.iii p. 297.
[137] BBCWAC 47796 (3 Mar. 1938).

relevant ISCM concert on 20 June it did agree to share the fee and to engage the necessary percussion players.[138] The skirmishing for engagements resumed between Mrs Kossar and the Music Department soon after this concert. This time Kossar adopted a 'superior attitude', in the hope of eliciting a promise of engagement.[139] Bartók had set aside some weeks in November 1938 for concerts, she claimed, but did not want a total of more than fifteen concerts in northern Europe. She therefore needed to know, in principle, whether the BBC was interested in being included in that limited number of engagements. A. Wynn in the Programme Contracts Department gave the regular response, of interest, but provided no definite promise.[140] When the programme schedules were drawn up the BBC could offer nothing in November, although it suggested that something might be possible in December.[141] Bartók immediately rejected any engagement at that time, however, as it would have been pointless to travel so far for a single appearance.[142]

In November 1938 Kossar tried again to gain a commitment, by informing the BBC that Bartók would be playing his Sonata for Two Pianos and Percussion in Paris on 27 February 1939 and could easily come on to London.[143] Wynn replied for the BBC:

We regret that we are unable to break our policy by bringing Mr. Bartók over from Paris in February or early March. If and when he and Mrs. Bartók are already in England on other business, we will be only too glad to consider it. It is only fair to state, however, that we have no Contemporary Concert in prospect in which the Sonata would fit, and its nature precludes its forming part of an ordinary programme.[144]

Bartók would not be made a special case, especially as his recent masterwork had been judged difficult to programme. As Julian Herbage noted some years later, the Bartóks' performance of the sonata in June 1938 had been found to be 'pretty stiff going even by those without any antipathy to contemporary music'.[145] From the BBC's viewpoint any broadcast which included this work would have needed not just careful placement but also a great quantity of ambient sweetening to disguise its 'bitter pill'.[146]

By May 1939 Dr P. Schiff in Paris had taken over Bartók's British representation, and began to try for engagements where Mrs Kossar had

---

[138] MTA BH 481; BBCWAC 47796 (15 Mar., 25 Mar. 1938).
[139] BBCWAC 47796 (2 July 1938).
[140] BBCWAC 47796 (13 July 1938).   [141] See KBKK pp. 213–14.
[142] KBKK p. 214.   [143] BBCWAC 47796 (17 Nov. 1938).
[144] Unpublished letter, Wynn (BBC) to Kossar, 9 Dec. 1938, BBCWAC 47796.
[145] Unpublished BBC memorandum, Herbage (ADM) to Lennox Berkeley (BBC), 12 Apr. 1943, BBCWAC 47796. As detailed in Chapter 6, below, the performance was actually warmly received by many London critics.
[146] On programming policies at this time see Basil Maine, The B.B.C. and its Audience (London, 1939), 52–3.

failed. His first suggestion was for a performance of the Sonata for Two Pianos and Percussion with the BBC in November or December 1939.[147] The standard non-committal reply came back from the BBC.[148] By return Schiff suggested a performance during the period 10–18 December. The BBC noted this.[149] In August Schiff wrote again suggesting a concert in early January 1940. The BBC, in reply, again noted this.[150]

Events of 3 September 1939 made any further correspondence pointless.

[147] BBCWAC 47796 (12 May 1939).
[148] BBCWAC 47796 (19 May 1939).
[149] BBCWAC 47796 (22 May, 30 May 1939).
[150] BBCWAC 47796 (16 Aug., 22 Aug. 1939).

6.

BARTÓK AT LARGE

While for nearly a decade after his visits of 1923 Bartók performed exclusively with the BBC, other British institutions and individuals were less obtrusively playing their parts in consolidating his reputation. Through the support of several music critics Bartók quickly assumed a high profile in reference and generalist music literature. A. Eaglefield Hull's *Dictionary of Modern Music and Musicians* of 1924 devoted a sizeable entry to Bartók and, moreover, through Bartók's authorship of numerous articles on Hungarian musical topics, made his views more accessible to the English reader.[1] In *Music: Classical, Romantic and Modern*, which appeared in 1927, just before Hull's death, he hailed Bartók as 'one of the six chief personalities in twentieth-century music', and provided a perceptive analysis of his techniques.[2] M.-D. Calvocoressi continued irrepressibly to bring forward Bartók's name in his journal articles, particularly those concerning opera, since he considered *Duke Bluebeard's Castle*, Op. 11 to be one of the few modern operas in which the music equalled the text 'in matters of compression and speed'.[3] His book of 1925 entitled *Musical Taste and how to Form it* was even noted in the Hungarian press for its favourable treatment of Liszt, Kodály, and Bartók.[4] Another supporter, Eric Blom, ensured that Bartók gained a substantial entry in the 1927 edition of *Grove's Dictionary of Music and Musicians*.[5] There, however, Blom further entrenched the tendentious argument put forward by Cecil Gray seven years before in the *Sackbut*: 'Bartók considers himself bound by no accepted rules, and even the logical procedures of his own making he changes deliberately with almost every new work that he writes. He has no fixed method of gaining his ends, and his outlook varies in accordance with the requirements of the moment.' Two years later in *Cobbett's Cyclopedic Survey of Chamber Music* Edwin Evans confronted a similar problem in making brief but meaningful generalizations about Bartók's methods.[6] While the first two string quartets yielded to his investigations, the two violin sonatas defied his exegetical skills: 'Here, such explanation as could

[1] (London, 1924), pp. 29–31. Many of the articles written by Bartók are listed in *BBE* p. 538. Bartók contributed also to the article on 'Harmony'.
[2] (London, 1927), pp. 224–8.
[3] 'Bartók's "Duke Blue Beard Castle" ', *MMR* 54 (1924), 36. See also *MB* 7 (1925), 172–3, 202–4.
[4] (London, 1925). See review, Anon., 'Bartók és Kodály angol dicsérete', *Zenei Szemle* (Budapest), 10. 1 (Nov. 1925), 23.
[5] 3rd edn., ed. H. C. Colles (London, 1927), i. 232–5.
[6] ed. Walter Willson Cobbett (London, 1929), i. 60–5.

find place in a dictionary ceases to have value. The writing is mostly pure expressionism, and to apprehend it is a matter, not of understanding, but of sensibility. For that reason, all argument is futile.'

New supporters emerged to promote Bartók's interests in fresh quarters. Dezső Rácz, at the Hungarian Consulate, initiated the earliest contacts with the BBC, while his friend the journalist Frank Whitaker harboured a more lofty scheme: the publication of the first full-length book about Bartók in any language. Whitaker had interviewed Bartók in Budapest during late 1925, and described their long discussions in a *Musical Times* article of March 1926.[7] He finely portrayed the antithesis in Bartók's character—the rugged aloofness of the public composer and performer, the gentle courtesy of the private man—before reporting on his activities in collecting folk-music and his views on modern music. Bartók recognized Stravinsky as 'the greatest of the moderns'; Schoenberg lacked the inspiration of Stravinsky, in his opinion, but still possessed a praiseworthy technical skill. More generally, Bartók stated that a great harmonic field still remained to be explored by contemporary composers, but that this exploration had not to be at the expense of melody. Whitaker's article was commended for its veracity by Bartók, as well as various members of the Hungarian press.[8] The composer was flattered by Whitaker's suggestion soon afterwards of a book about him and gave his enthusiastic support to the project. On 16 August 1926 Whitaker accordingly signed a contract with Oxford University Press.[9] For the time, the project was certainly ambitious. Condition 4 of the agreement stated that the work was to be translated from English into German and Hungarian. Whitaker started to fashion his material into four chapters: the Man, Folksong, the Music itself, and the Problem of the New Music.[10] But, despite help from Bartók, Calvocoressi, and Rácz, he made only slow progress. Eventually in 1929 Whitaker sent a draft of his first chapter to Bartók, who made detailed corrections and revisions to it.[11] Whitaker promised to send a further chapter later in the year, but this promise was never fulfilled. The project died. Something of Whitaker's approach and style can, nevertheless, be gleaned from this draft of his opening paragraph:

It is fitting that Hungary, that land of turbulent spirits, should have produced Béla Bartók. You detect the type at first glance—the lean, tense frame, the flying white hair that only some fugitive studio photographer has succeeded in taming, the wild, bright eye that no one will ever tame, all tell of the born fighter—the man whose faith in himself is unquestionable, who is not indifferent to the world's opinion, but who will not go an inch out of his way to woo it. One of Bartók's publishers once

[7] 'A Visit to Béla Bartók', *MT* 67 (1926), 220–3.
[8] BL Add. MS 51023 B/131.
[9] Ibid.; Copy of Agreement, dated 16 Aug. 1926, BL Add. MS 51023 B/118–20.
[10] See synopsis, BL Add. MS 51023 B/1.
[11] See MTA BH 1970.

said to me sadly: 'If only Bartók would write another string quartet like the first we would all fall on his neck. We have told him so many times, but he won't do it.' How often one hears this sort of thing of any superior who makes progress, as if inspiration were only another name of duplication. If Bartók were capable of writing another quartet like the first he would believe it was frustrating his art to attempt it.[12]

On the concert-platform Bartók's works continued to receive a hearing, thanks to the support of George Woodhouse, who invited the Budapest String Quartet to perform Bartók's First Quartet, Op. 7 at his London school on 19 January 1925, and Sir Henry Wood, who gave the British première of the *Dance Suite* (1923) at the Queen's Hall on 20 August of that year. This latter performance started a small-scale wave of popularity for Bartók's music. By November Curwen and Sons, the British agent for Bartók's publisher, Universal Edition, felt justified in taking out a full-page advertisement for Bartók's scores in the *Musical News and Herald*, headed by the *Dance Suite*, 'to be heard in 60 cities next season'.[13] The work received many performances in London and was not overlooked in provincial cities. On 2 March 1926 Adrian Boult programmed it for a broadcast Birmingham City Orchestra concert, drawing a less enthusiastic response than that accorded in the capital. While critical opinions were divided,[14] the audience was close to united in its judgement—at least according to one correspondent:

The present writer has exchanged opinions with 173 members of the audience at the Symphony Concert on March 2nd; and all but nine have said what the critic of the 'Times' said after a London performance of the Suite on February 25th: 'We shall slip away before it, next time it occurs. It is time that some protest was registered against glissandos on the trombones, even if every other atrocity is admitted to the concert room.'[15]

Boult was undeterred. Cheekily he added at the conclusion of the published programme for 23 March: 'In response to many requests the Dance Suite [of] Béla Bartók performed at the last Concert will be repeated.'[16] According to contemporary reports and Boult's own reminiscences the work was better received at its second hearing.[17] Even in Manchester the conservative conductor of the Hallé, Hamilton Harty, allowed the suite to be performed.[18] Although the most popular, the *Dance Suite* was not the only work to bring Bartók's name before the musical public at this time. His

[12] BL Add. MS 51023 B/33.   [13] *MNH* 7 Nov. 1925, p. 436.
[14] See reviews: *Birmingham Gazette*, 3 Mar. 1926; *Birmingham Post*, 4 Mar. 1926; *Birmingham Gazette*, 4 Mar. 1926.
[15] Letter to the Editor, *Midland Musician*, 1 (1926), 160.
[16] Programme in János Demény private collection (Budapest).
[17] Anon., 'The City Orchestra', *Birmingham Post*, 24 Mar. 1926; reminiscences of Sir Adrian Boult, in János Demény private collection (Budapest).
[18] See Michael Kennedy, *The Hallé Tradition* (Manchester, 1960), 214, 224, 253–4.

Second String Quartet was performed in London on 20 November 1925 by the Pro Arte Quartet, and broadcast by the Hungarian Quartet on 19 February 1926. In 1926 plans were even drawn up for a performance of Bartók's ballet *The Wooden Prince*, Op. 13, although they were not carried through.[19]

During the 1930s, as engagements with the BBC became harder to obtain, Bartók had need to exploit the community awareness generated by his supporters in the preceding decade. He was increasingly thrown back upon his friends and associates for offers of performance and had to rely more on other performers to provide the living experience of his works to British audiences. It was through the violinist József Szigeti that Bartók secured his first British engagement outside the BBC in over eight years: a joint recital with Szigeti in the Oxford Town Hall on 4 February 1932. This was a humble re-emergence into 'private enterprise' performance. As part of a standard subscription series, the programme could not feature Bartók's music too extensively. It began with a Mozart violin sonata and Bach solo violin sonata, and only came to Bartók's Second Violin Sonata (1922) and a series of his shorter piano pieces in the second half. Neither the fee nor the agency commission rates were favourable,[20] especially as Bartók had come to Britain solely for this concert. It was work in a new field, however, and the efforts of both artists were appreciated: 'We may call ourselves lucky that Mr. Bartók and Mr. Szigeti are both "artists with an international reputation" ', commented the student paper the *Cherwell*, 'for their exclusion under the Home Office protection scheme would have deprived us of one of the most remarkable events of this term's concert season.'[21] The few reviews of the concert in local papers steered clear of contentious value-judgements, praising the performers, particularly Szigeti, for their fine playing, and finding much of interest in Bartók's sonata. Only the piano pieces gained some adverse comments. About 'The Night's Music' from *Out of Doors* (1926), Trevor Harvey commented in the *Oxford Mail*: 'Its fault, perhaps, was that the night was rather long and uneventful, the composer seeming to attempt to reconcile the beauty of a summer's evening with the length of one in the winter.' The *Cherwell*'s critic found an opposite problem with some of the piano pieces: they appeared too small and inconsequential in the company of major masterpieces of the violin repertory, and were more encore than programme pieces.

Later in February 1932, after several weeks back on the Continent, Bartók went north of The Border for a concert in Glasgow before the Active

[19] See András Benkő, 'Romániában megjelent Bartók-interjúk', in *Bartók-dolgozatok 1981*, ed. Ferenc László (Bucharest, 1982), 310.

[20] See MTA BA-B 3477/213.

[21] Reviews mentioned: A. A., 'Oxford Subscription Concert', *Cherwell*, 13 Feb. 1932, p. 91; Trevor Harvey 'Béla Bartók Plays at an Oxford Concert', *Oxford Mail*, 5 Feb. 1932. For a further review see *Oxford Times*, 5 Feb. 1932.

Society for the Propagation of Contemporary Music. While his music had been much played in England, and even heard in Wales, it had received less attention in Scotland. A performance of *Two Portraits*, Op. 5 was given by a Glasgow orchestra in 1927 or 1928,[22] but it was not until 13 December 1929 that a concert of Hungarian music was mounted, by Erik Chisholm's newly formed Active Society.[23] Chisholm had only recently returned to his native Glasgow after several years in Canada, where he had spent considerable time promoting the music of Bartók and Schoenberg. This performance of 1929 was adventurous, if not a little bizarre. After numerous Bartók and Kodály piano pieces, played by Chisholm and the Society's Secretary, Patrick Shannon, the two combined for a performance of Bartók's First Piano Concerto (1926). Chisholm played the solo piano part, while Shannon did his best to represent the orchestra—on the piano for the outer movements, but moving to the organ for the slow middle movement. About the concerto's performance the *Glasgow Herald*'s critic wrote:

The effect of Bartók's very individual scoring was naturally missed, but something of its quality could be imagined from the nature of the music for second piano. Certainly the performance was worth while, even without the orchestra, for the concerto is a notable work, very distinctive in its material and treatment, and quite unusual in the power and persistence of its rhythmic effects. We may have to wait a long time for a performance with orchestra.[24]

Even those critics who found the lack of 'agreeable consonances' distressing recognized the importance of these first Scottish performances of works by such acclaimed Hungarian composers.

Chisholm, then aged only 25, was much encouraged by the reception of this concert, and ventured to write a short letter to the Master himself telling of the recital and enclosing a few reviews.[25] Moreover, he told of his plans to propagate Bartók's music widely, through repeat performances of the concert in Edinburgh, Stirling, St Andrews, and Aberdeen. Bartók did not reply. But Chisholm's enthusiasm was not easily dampened. The Edinburgh concert took place on 8 February 1930, leading to a second letter, in which Chisholm professed 'unbounded respect and admiration' for Bartók's First Piano Concerto (which was to be performed by Bartók over the radio later that week).[26] Bartók did not reply, necessitating a third letter from Chisholm in June. It began:

The above Active Society (consisting of several energetic young musicians) are out for [*sic*] active propaganda for contemporary composers. We have already last

[22] See *DB.iii* p. 153.
[23] Programme, MTA BH 2049/298. The concert took place in St Matthew's Church of Scotland, Bath Street, Charing Cross.
[24] Anon., 'Contemporary Music Recital', *Glasgow Herald*, 14 Dec. 1929.
[25] *DB.iii* p. 149.   [26] MTA BH 285.

season given 2 Recitals entirely devoted to your works, and there are now a large
number of enthusiastic Bartókians in Scotland thanks to our activities. . . . I know the
London B.B.C. engage you from time to time to perform at their Contemporary
Music Recitals, and I wonder if you would consider paying us a visit in
Scotland . . .[27]

Chisholm could not match the BBC's fees, but offered travelling expenses,
accommodation, and any profits generated by the concert. Touched by such
persistence Bartok did reply, expressing interest in the idea.[28] Chisholm
was gratified and soon furthered his exchange with Bartók:

Of course I want you to come to Scotland more than anything else—if this should
prove impossible for this year then next season after a year's trading our Society may
surely be able to offer you a fee. As it is, if one of these organisations I have written
[sic] can arrange an appearance (fools if they cannot) I shall do everything in my
power to make your recital a success, & should hand to you the entire proceeds
drawing of the Recital.[29]

A recital could not conveniently be organized for the coming season,
although Bartók did become a member and later Vice-President of the
Society.[30] In the Society's 1930–1 season, none the less, local musicians
performed several Bartók chamber works and Chisholm even intended to
mount Bartók's opera, but found the financial outlay for an orchestra too
onerous for the Society in a time of such economic depression.[31] During the
summer of 1931 Chisholm negotiated with Bartók's Viennese agent for an
engagement in early 1932, eventually securing a commitment for 29
February.[32] In the meantime, on 5 November 1931, the Kolisch Quartet
had performed Bartók's Third String Quartet (1927) in Glasgow, further
enriching local appreciation of his music.

At 8.35 p.m. on 28 February 1932 Bartók arrived in Glasgow aboard
'The Flying Scotsman', to a welcome from a large party including the well-
known art collector and Honorary Hungarian Consul, Sir William Burrell,
members of the Active Society, and press photographers.[33] As Diana
Brodie, the Society's Secretary and Chisholm's wife, recalled, no one present
knew Bartók by sight. But when the passengers alighted all doubts were
dispelled: 'There was only one Béla Bartók! A small white-haired man,
wearing a black Homburg hat, thick black coat with a heavy astrakhan
collar, and armed with a music case in one hand and an umbrella in the

[27] Unpublished letter, Erik Chisholm to Bartók, 3 June 1930, in English, MTA BH 287.
[28] Bartók's letter is lost.
[29] Letter, Erik Chisholm to Bartók, 30 June 1930, in English, DB.iii p. 153.
[30] MTA BH 289, 291.
[31] Prospectus, MTA BH 2400/733; MTA BH 290.
[32] See unpublished correspondence: MTA BH 811–12; MTA BH 291; MTA BA-B 3477/65
(Plate 8).   [33] See Plate 9.

other. Who, I wondered, had forewarned him about Glasgow's weather?'[34]
The following day was hectic, with long hours of rehearsal for Bartók in
solo and chamber roles before the concert that evening in the Stevenson
Hall. Its long all-Bartók programme is reproduced below:[35]

I.   a.   Second Elegy from Op. 8*b*
       b.   Three Burlesques Op. 8*c*
BÉLA BARTÓK—Piano

II.   Five *Village Scenes* (1924) for voice and piano
ANGELA PALLAS—Voice BÉLA BARTÓK—Piano

III.   a.   Romanian Christmas Songs (1915), Set I
       b.   Suite Op. 14
       c.   *Allegro barbaro* (1911)
BÉLA BARTÓK—Piano

IV.   a.   *Lassú*, from the First Violin Rhapsody (1928)
       b.   Sonatina (1915) (trans. Gertler)
       c.   Romanian Folk Dances (1915) (trans. Székely)
BESSIE SPENCE—Violin BÉLA BARTÓK—Piano

INTERVAL

V.   Four songs from Eight Hungarian Folksongs (1907–17)
ANGELA PALLAS—Voice BÉLA BARTÓK—Piano

VI.   a.   'Preludio, all'ungherese', from Nine Little Piano Pieces (1926)
       b.   'The Night's Music', from *Out of Doors* (1926)
       c.   'With Drums and Pipes', from *Out of Doors* (1926)
       d.   First Romanian Dance, from Op. 8*a*
BÉLA BARTÓK—Piano

For Bartók the evening afforded a most pleasant introduction to a
Scottish audience. According to one columnist the concert had attracted a
large body of 'music lovers from all parts of the West of Scotland'.[36] Many
of these had already acquired some knowledge of Bartók's music through
the Society's concerts or BBC broadcasts. Bartók's vocal and string
associates, of only that day's acquaintance, performed most competently

[34] Unpublished paper, 'Béla Bartók' [1964], by Erik Chisholm, pp. 9–10, now in the
possession of Chisholm's second wife, Mrs Lillias M. Forbes (St Andrews). This was Bartók's
usual travelling attire in the colder months of the year.
[35] MTA BH 2400/75. Errors in dating and citation have been corrected.
[36] Reviews mentioned: Anon., 'Mr. Béla Bartók in Glasgow', *Scotsman*, 1 Mar. 1932;
W. S. G., 'Folk Music From the Near East', *Evening News* (Glasgow), 1 Mar. 1932; Anon.,
'Music in Scotland', *MO* 55 (1931–2), 610; H. K. W., 'Béla Bartók Stirs Large Audience',
Bulletin and Scots Pictorial, 1 Mar. 1932; Montague Smith 'Active Society', *Evening Citizen*
(Glasgow), 1 Mar. 1932.

and elicited a generous applause after each item. That applause flowed into the review columns of the newspapers, which displayed a politeness and appreciation long since abandoned by the English critics. Even Bartók must have been surprised at the praise of his piano playing, which rather eclipsed the compliments about his compositional skills. Seldom these days did he read, as in the *Evening News*: 'Mr. Bartók is an excellent pianist, full of vitality and a subtle sense of tone and colour effects.' Nor, as in *Musical Opinion*: 'He has amazing independence of fingers which makes his playing of heavy chordal passages most interesting. Passages that would sound over-burdened at the hands of the ordinary pianist become, under Bartók, glowing with life and full of colour.' The mellow nature of the programme, with a considerable number of early or overtly folk-influenced pieces, partly explained this level of praise—Bartók had not sought to confront his new audience with the monumental works of his output.

Evaluations of the works from a compositional viewpoint were mostly avoided. The critic from the *Bulletin and Scots Pictorial* did, however, find Bartók's 'outstanding abilities and originality' more clearly demonstrated in the works of later rather than earlier years, and suggested that these more recent works would prove an important influence on mainstream European culture. Driving rhythms and folk influences were mentioned in a general context by many of the writers, but were not subjected to any detailed probing. Montague Smith in the *Evening Citizen* broadened his critical perspective to compare Bartók with the other two famous names of contemporary music, Schoenberg and Stravinsky. He found Bartók 'far more direct, less abstract than the former, and less dry and crafty than the latter, and he is more human than either'. And Bartók's report on the day? Before going to bed he penned this short postcard:

Dear Mother and Aunt Irma,

Today was a very busy day for me here: I rehearsed with the singer and violinist for approximately four hours, and practised a little myself. The concert was in the evening. Everything went well. Yesterday evening when I arrived at the station they took photos for the papers. It's a long time since I ate as bad a lunch as today—and at an exclusive hotel. The Scots don't seem to be able to cook, only to play the bagpipes. Many kisses, Béla.[37]

While in Glasgow Bartók stayed with the Chisholms. They found that his natural reserve could quickly be overcome and that when he started to relax he revealed the forceful personality behind that quiet, unassuming façade.[38] Chisholm spoke about Scottish folk-music and his pioneering research in the field. This encouraged Bartók to listen to his folk-recordings

---

[37] Postcard, Bartók to his mother, 29 Feb. 1932, *BBcl* p. 525.

[38] The information in this paragraph is drawn from Chisholm, 'Béla Bartók', pp. 10–13. Despite Diana Chisholm's statement that Bartók was 'really badly off financially', he was probably more affluent in the 1930s than at any other time in his life.

for hours on end, and even, on the following day, to buy a tartan rug, chanter, and all available Piobaireachd music. Bartók's enthusiasm went as far as to arrange for a well-known bagpipe player to perform for him. He was thoroughly pleased with his new experiences. During this visit Chisholm's wife noted closely some of Bartók's more personal features:

He made a great fuss of our baby daughter Morag and seemed to be extremely fond of children, yet I felt he had built an invisible barrier of defence for himself against the outside world. We do know that he was really badly off financially, and that apart from his heavy overcoat, which was beautifully warm and looked new, his suits though well-tailored and well pressed were equally well-worn, his shirts too were frayed at the cuffs and collars; altogether he gave one the impression of 'putting a face' on things generally, and being harassed by some secret worry. The face of a pathetic little man—but an intensely proud one who was also a musical genius.

When served a Scottish high tea and asked if he wanted jam, Bartók replied: 'No, thank you', but then added, 'is it manufactured or home-made?' On learning that it was home-made, Mrs Chisholm recalled, he changed his mind: 'Then I'll have some. Home-made jam has character and taste, bought jam neither.' When rehearsing for the concert Bartók pointed out to Chisholm a very difficult part in the score, suggesting that he, as page-turner, should play some of the notes in the bass to 'help out'. Although surprised, Chisholm agreed. In the actual performance Chisholm was about to 'help out', when Bartók gave him a 'flashing impish grin', and played the section solo in quite brilliant fashion. In rehearsal Chisholm also noted Bartók's fascination with the duration of pieces. Using a stop-watch he would time each piece to the second, shaking his head in reproof if the watch produced anything other than the expected result.

Without doubt Bartók enjoyed his short visit to Scotland, for he agreed to come again late in the following year for a similarly small fee.[39] Because of troubles in getting a firm commitment from the BBC for his London engagements, the date of 2 November 1933 was only finalized at a late stage after much correspondence.[40] On his arrival in Glasgow Bartók was again met by an official delegation, including press photographers, before going to stay with the Chisholms.[41] The concert, a solo piano recital, was the first of the Active Society's new season, and took place in the new venue of St Andrew's (Berkeley) Hall. After beginning with several short works by Purcell and a number of seventeenth-century Italian masters, Bartók played a selection of piano pieces by Kodály, his own Sonata (1926), and a collection of his smaller piano compositions. Instead of the usual programme-

[39] Ibid. 4. Bartók agreed to a fee of £15.
[40] See: unpublished letters, Bartók to Erik Chisholm, 13 Aug., 29 Aug., and 13 Sept. 1933, in English, ibid. 4–4B; BBCWAC 47796 (12 Sept., 20 Sept. 1933); BBL p. 222.
[41] Cf. Plate 10.

notes, the audience received eight Roneoed pages of handwritten analyses of the various works, accompanied by nearly forty musical examples.[42] Although thirty years later Chisholm claimed that these notes had been written by Bartók,[43] they were probably his own elaboration of some of Bartók's ideas.

As in 1932, the concert was well received by about 300 Society members and supporters. 'Bartók Delights Large Audience' was the type of headline rarely read by the composer.[44] Perhaps because of the slightly heavier nature of the programme, the critics expressed a few more reservations than in the previous year, concerning both Bartók's pianism and composition. The opening bracket of early music received the most united criticism. Despite the apparent authorships of Purcell, Marcello, and others, this music came across as pure Bartók. *Musical Opinion*'s writer found Bartókian traits in every line; according to the *Glasgow Weekly Herald* Bartók had 'crystalised them, percussioned them and shed over them the white light of his own incandescent personality'. As Erik Chisholm has recorded, Bartók had discussed such criticisms as these during the morning rehearsal for the concert:

Bartók turned to me saying: 'You know, Mr. Chisholm, that whenever I play these transcriptions, the critics always complain that I have made considerable modifications in the originals. As a matter of fact, I have not altered a single note.' Although I didn't say so at the time, I could see why this mistake had been made, for Bartók played this music in his own dynamic, rhythmically arresting fashion, so that, even if all the notes were the same, the music *sounded* as though Bartók had altered it.[45]

For the following Kodály pieces, standard works in Bartók's repertory for many years, the critics accorded only praise, finding Bartók the ideal exponent of such music because of his command of tone gradation, rubato, and pedalling. The final selection of his own compositions, too, was well received, with the exception of the largest work, the Sonata (1926). While rhythmic momentum carried the listener through the sonata's first movement, claimed the *Bulletin*, the slow second movement palled away because of its unattractive idiom and became forbidding. For the conservative critic from the *Glasgow Weekly Herald*, the breaking down in this sonata of the barriers of barlines and restrictions of key signatures only led to aural confusion. Not all found such problems with the work, however. Chisholm recalled:

---

[42] Programme, MTA BH 2400/89.   [43] Chisholm, 'Béla Bartók', pp. 5–6.
[44] Reviews mentioned: H. K. W., 'Bartók Delights Large Audience', *Bulletin and Scots Pictorial*, 3 Nov. 1933; Anon., 'Music in Scotland', *MO* 57 (1933–4), 232–3; 'Maestro', 'Music and Musicians', *Glasgow Weekly Herald*, 11 Nov. 1933; Anon., 'Piano Recital', *Glasgow Herald*, 3 Nov. 1933.   [45] Chisholm, 'Béla Bartók', p. 6.

I have never heard anything to equal the rhythmic intensity, the sheer percussive vitality, the dash and abandon, the actual physical reality (some critics called it brutality) of the sound content in Bartók's playing of the first and last movements. . . . When he played, the legs of the piano seemed to be twitching in an effort to join in this animalistic, choreographic, Pan-worship rite.[46]

Despite this variety of opinions, Bartók's own compositions were generally agreed to be the most gripping on the evening's programme. To 'Maestro' from the *Glasgow Weekly Herald* Bartók's 'music of the mind' was best described in terms of colours: 'He has an amazing faculty of keeping three or four contrasted tone qualities going at one and the same time, each quite clearly discernible, but the tone pictures that result are compounded of whites and greys, blues and heliotropes—the red and yellow elements that give the warmth are invariably absent.'

Bartók was again pleased with his reception in Glasgow. Although he could gain three times his Scottish remuneration for an hour with the BBC, it was difficult to promote his music so actively and directly in London, and be so appreciated. While in Glasgow Bartók discussed with Chisholm some of his recent works—the Second Piano Concerto (1931) and *Forty-four Duos* (1931) for violins—copies of which he had in his music case.[47] Sitting in Bartók's guest bedroom they talked about the differences in piano writing between his two concertos and the ways of playing some of the chords. Chisholm wondered at Bartók's use of tone-clusters and learnt of the adoption of the technique from the American composer Henry Cowell, when they had both been staying in London during December 1923. Bartók also spoke of his plans for the *Mikrokosmos* series of graded piano compositions, then in its early stages. After a short country excursion with the Chisholms,[48] Bartók left for his BBC engagements in London. Although a correspondence was maintained until 1938, no further concerts were arranged.[49]

As his fortunes with the BBC continued to decline Bartók re-established links with the British Office of the International Society for Contemporary Music and with the regional associations originally affiliated with the British Music Society. Two concerts were planned with the Liverpool Music Society, in 1936 and 1938, although only the earlier concert took place.[50] This Liverpool performance, on 16 January 1936, involved collaboration with the violinist Zoltán Székely. It was a rushed visit, scheduled between a BBC rehearsal in Birmingham and two concerts there. In the Rushworth Hall the duo presented a mixed programme of Mozart, Bartók, and Ravel violin sonatas, as well as Bartók's *Romanian Folk Dances* and First Violin Rhapsody. After the more informed reviews in London, and even Glasgow, the Liverpool criticisms appear rather naïve. The most substantial review

[46] Ibid. 6–7.   [47] Ibid. 2–3.   [48] *BBcl* p. 541.
[49] MTA BH 292–3.   [50] KBKK p. 213.

was published in the *Liverpool Daily Post*, penned by A. K. Holland, who had written one of the most interesting and honest of the reviews on Bartók's previous visit to the city in 1922.[51] Holland had not progressed, however, in his understanding of Bartók's music, nor of the broader trends in contemporary music. While recognizing Bartók's acquisition of a truly international status since his first visit, Holland confessed that he found Bartók's music just as troublesome. He accused the composer of being too abstruse in his Second Violin Sonata (1922)—indeed, of leaving him, as a listener, 'metaphorically deaf'. Holland could at least discern the merit in Bartók's works of smaller form—the dances and rhapsody—but felt compelled to question his musicianship in interpreting the Mozart and Ravel sonatas. Although in good taste, consistent, and individual, Bartók's playing was, he believed, too dry and unromantic. 'He is still a rather incalculable force in modern music', Holland concluded. For *Musical Opinion*'s local correspondent matters were simpler. Bartók's activity was 'ugly', and 'gritty', and 'somewhat aggressive'.[52]

In the following year, on 9 February, Bartók performed for the ISCM at London's Cowdray Hall. Although he had not given a concert for the British branch of the organization since May 1923, Bartók had meanwhile taken part in numerous ISCM Festivals across Europe and communicated periodically with the Society's founding President, Edward J. Dent, about the formation of a Hungarian section of the Society.[53] Because of the BBC's lack of interest in unveiling to the world the first selection of *Mikrokosmos* piano pieces—the most widely disseminated of all his compositions today— this honour had fallen to the ISCM: 'a great score for this small and go-ahead body', as the *Evening News* commented.[54] The concert also provided another opportunity of collaboration with Zoltán Székely following their BBC venture during the previous week. Accordingly, the evening began with the First Violin Sonata (1921), ended with the Second Violin Rhapsody (1928), and featured twenty-seven *Mikrokosmos* pieces in between.

Rehearing this violin sonata after several years, some critics compared its current reception with its initial impact during Bartók's first post-war visit. The *Musical Times*'s critic, for instance, recalled how its harsh sounds had then 'bewitched the audience into being so many mental porcupines with their quills out'.[55] By 1937 the sonata was much more intelligible, even if it

---

[51] Cf. A. K. H., 'Liverpool Music Society: Béla Bartók's Recital', *Liverpool Daily Post*, 17 Jan. 1936, and A. K. H., 'M. Béla Bartók', ibid., 31 Mar. 1922.

[52] Anon., 'Music in Liverpool', *MO* 59 (1935–6), 422.

[53] See, e.g., MTA BH 1478–9.

[54] W. McN. [William McNaught], 'Béla Bartók Plays New Piano Pieces', *Evening News*, 10 Feb. 1937.

[55] Reviews mentioned: Anon., 'London Concerts', *MT* 78 (1937), 267; R. C., 'Béla Bartók in London', *DT* 10 Feb. 1937; Anon., 'Concerts of the Week', *Observer*, 14 Feb. 1937; Anon., 'Contemporary Music Centre', *The Times*, 11 Feb. 1937. Other reviews appeared in *DM* 10 Feb. 1937; *Evening News*, 10 Feb. 1937; *Magyarság*, 17 Feb. 1937.

still came across as a 'long, biting work'. In the *Daily Telegraph* Richard
Capell recognized a good performance revealing the 'undeniable poetic
quality of the music'. Even the most anti-Bartók paper, the *Observer*, could
compromise its stand sufficiently to acknowledge 'genuine music' and
'moments of rare feeling in it behind the unintelligible cacophony'. The
central feature of the concert, the selection of *Mikrokosmos* pieces, elicited
more high-flown prose than solid sentences of evaluation. 'Percussive steps
to an Hungarian Parnassus', offered Capell. For the *Observer*'s anonymous
critic, 'the piano pieces are science, as interesting no doubt to write (and, for
those who can, to solve) as chronograms and palindromes, and as devoid of
musical vitamins'. *The Times*, however, examined these pieces more
penetratingly. Its critic identified much 'linear counterpoint' in the pieces,
where 'the parts do not fit, but merely set up a tension one against the
other'. This could be a recipe for disaster, the critic considered, but Bartók
usually avoided such catastrophe by finally resolving to some kind of
concord. The *Musical Times*'s writer found these pieces suggestive of the
cleanness of steel. All superfluities had been stripped away, while the special
difficulty of each individual piece was presented and then conquered. Hence
the titles, such as 'Fourths', 'Major Sevenths', 'Syncopation'. Only
occasionally, as in 'From the Diary of a Fly', did an element of grim humour
creep in. If for children, this critic suggested, then these pieces must be for
the offspring of an iron age.

During this London visit of February 1937 Bartók again stayed with the
Wilsons in South Kensington.[56] By 1937 Duncan Wilson was in his early
sixties; he was His Majesty's Chief Inspector of Factories, and in line for a
knighthood the following year. This steady civil service background had
probably inspired Bartók to seek help from the couple on various insurance
and, probably, income tax matters.[57] It appears that Bartók repaid these
services musically, by giving lessons to Duncan, a keen amateur pianist and
enthusiast for Bartók's more manageable works.[58] For Freda's amateur
cello playing he had less patience, as Sybil Eaton, a close friend of the
couple, recalls.[59] When Freda once announced that she was going to
practise, Bartók asked if there were any ear-plugs to hand! To the Wilsons'
friends he gave the impression of being kindly (he brought them a generous
present each time he stayed), but rather pedantic. When Bartók spoke to his
hosts his English was most precise, but not fluent. Sybil Eaton remembers a
great search for the word 'antlers' (from the *Cantata Profana* story), Bartók
being totally unable to accept any English synonym.

[56] *BBcl* p. 572.   [57] See MTA BH 2003; MTA BH D-II.32/2, 35–7 Számlák.
[58] See MTA BH 2002; MTA BA-B 3477/233. Mention is made of Duncan Wilson's amateur
performing in *The Oxford and Cambridge Musical Club*, ed. Graham Thorne (London, 1979),
18. Wilson had been a founding member of the club in 1900. Bartók's letters to the Wilsons
were destroyed, along with all their personal papers, upon Lady Wilson's death in 1977.
[59] Interview, Sybil Eaton (South Kensington) with the author, 10 Feb. 1985.

On Bartók's final visit to Britain, in June 1938 for the ISCM Festival, he again hoped to stay with the Wilsons and to introduce to them his wife, Ditta, who was accompanying him to Britain for the first time.[60] Such a stay was not possible, however, and the Bartóks were accommodated by the Meighar-Lovetts, who lived in Cadogan Square, Chelsea. Staying in a new place and ever the wide-eyed child when confronted by real affluence, Bartók wrote home to his mother:

As I had already suspected, our lodgings in London are in a very stately home. All its furnishings are exquisite, only the very best. I don't even know how many rooms there are in the house. But I do know that there are five domestic staff: cook, housemaid, manservant, kitchen girl, nanny. The owner was not at home when we arrived (it turned out later that they had to go to a race meeting). They returned for the evening, and of course were most gracious and considerate. They even speak a little German. The manservant, though, is 'Schwyzerdütch' [sic], that is, Swiss German. We have our own private bathroom etc. etc.—but—one was required to dress for dinner![61]

In the Festival the Bartók duo performed his Sonata for Two Pianos and Percussion (1937), assisted by the percussionists Samuel Geldard and William Bradshaw. This work, the result of a commission from the ISCM branch in Basle, had been premièred successfully there during January, and performed less satisfactorily in Luxembourg when the Bartóks were on their way to Britain. Coming at the conclusion of the fourth Festival concert, on 20 June, the composition's London première was hailed by the critics as the most significant of the evening, or even of the Festival up to that point.[62] It shared the programme with such works as a Krenek cantata, *Variations for Strings* by Benjamin Britten, and *Two Choral Songs* by the Australian composer Peggy Glanville-Hicks. Among the critics, Richard Capell found in the Sonata music of a rare order, uncanny and demonic, with a poetry of sound similar to that in the Second Piano Concerto. For Scott Goddard in the *News Chronicle* the work was more futuristic than contemporary. He continued: 'It was so clearly constructed and functionally so well adapted to the medium that it was completely convincing even at a first hearing.' Alan Frank, in the *Musical Times*, along with *The Times*'s critic, also found the composition singularly successful and thoroughly gripping in its effect from start to finish. By exploitation of tonal affinities between the pianos, timpani, and xylophone, in particular, Bartók's 'first-class mind' had been led into great originality of thought, stated *The Times*. Despite under seven hours of rehearsals, Bartók was prepared to agree with these reviewers that

[60] MTA BH 1483.

[61] Letter, Bartók to his mother, 16 June 1938, *BBcl* pp. 588–9.

[62] Reviews mentioned: Scott Goddard, 'Contemporary Music', *News Chronicle*, 21 June 1938; Richard Capell, 'Béla Bartók and Others', *DT* 21 June 1938; Alan Frank, 'The I.S.C.M. London Festival', *MT* 79 (1938), 536; Anon., 'International Festival', *The Times*, 21 June 1938.

the performance had been a success.[63] As pianists he felt that they had played better, and more freely, than in Basle. The British timpanist, too, had been at least equal to his Basle counterpart, although the second percussionist had been somewhat less reliable.

While in London Bartók also took part in a function at the Boosey and Hawkes Organ Studio in Regent Street. His performance of fifteen short pieces from *Mikrokosmos* at this informal concert on the afternoon of 20 June signified publicly his new links with this publishing house. As soon as the German *Anschluss* with Austria had been proclaimed on 13 March 1938, Ralph Hawkes had realized that Bartók's Viennese publisher, Universal Edition, would soon be 'Nazified'.[64] Hawkes flew to Budapest to negotiate with Bartók and Kodály, hoping to gain sole publishing rights to their works. While Kodály demurred, Bartók immediately agreed with Hawkes's proposals, also consenting to change his copyright affiliation from the Viennese AKM organization to the English-based Performing Right Society.[65] Bartók's short studio performance, in the company of renditions by other performer–composers, Nikolay Lopatnikoff, Benjamin Britten, and John Ireland, was designed to popularize one of the more substantial Bartók works which Boosey and Hawkes intended to publish. By being placed at a central time within the week of the ISCM Festival, the concert drew a good audience and several reviews. The *Musical Times*'s critic best summed up the common opinion:

A group of piano pieces from Bartók's 'Mikrokosmos' made the impression of terrific energy that everything connected with him usually does. He 'impacted' into the room (there is no other word to describe the mental effect of his entrance, though his actions were quite quiet) and proceeded to play all sorts of little sketches, each one as if cut by a lapidary's tools, instead of written by a pen. Then he vanished again, leaving a wake of applause.[66]

These days in London with his wife were, to all appearances, happy, sunny ones, full of concerts and the many social events associated with the Festival. Bartók attended a performance of John Blow's opera *Venus and Adonis* at the Royal College of Music on 19 June, and also a concert of English folksongs and dances at Cecil Sharp House on 22 June. Ralph Hawkes invited the couple to dinner,[67] and the Mayers, who had been so hospitable back in the 1920s, threw a party, which Bartók and his wife attended. During this soirée, Gerald Abraham recalls that Bartók played several of the *Mikrokosmos* pieces.[68] It was possibly at this reception also

---

[63] *BBlev*.v pp. 592–3.

[64] Ralph Hawkes, 'Béla Bartók: A Recollection by his Publisher', in *Béla Bartók: A Memorial Review* (New York, 1950), 14.

[65] Ibid. 17. See also Ernő Balogh, 'Bartók in America', *Long Player* 2. 10 (1953), 4.

[66] Anon., 'London Concerts', *MT* 79 (1938), 536. See other reviews: *The Times*, 22 June 1938; *MO* 61 (1937–8), 856–7.   [67] MTA BH 133.

[68] 'Bartók and England', in *Bartók Studies*, ed. Todd Crow (Detroit, 1976), 166.

that both husband and wife played and Bartók started unexpectedly to improvise. In her novel *Béla Bartók's Last Years: The Naked Face of Genius*, Agatha Fassett recorded Ditta's memory of the incident in this way:

'I was playing a long solo part when all at once I felt as if a strange hand were touching my shoulder, and then, like a hallucination, a few very short and heart-breaking chords were floating toward me. Even when the sounds grew stronger, turning into wild, human voices, I still didn't know where they were coming from. But after the plaintive piercing last tones, so much more like a flute or a violin, had subsided, I realized that they were coming from Béla's piano. However, after a while, everything fell into order again and remained so to the end.'

'It was weeks later that I mentioned it to him. "Really, did you notice it?", he said, suddenly interested. But when I asked him what made him do it he shrugged his shoulders. "I don't know," he said. "I had a vision, for a second, that that very moment was my last, the last standstill, before everything should get lost in chaos and panic, and I felt I had to hold on to it." '[69]

From his youth Bartók had been acutely sensitive to the fluctuations of tension in Europe. In November 1937 he discerned dangerous directions in Hungarian Government policy and concluded that his manuscripts needed to be housed outside the country for safety's sake.[70] By June 1938 he had already dispatched several consignments to his friend Annie Müller-Widmann in Basle.[71] He recorded his true feelings in May of that year in the album of Ákos Weress, a collection of the three most profound wishes of famous artists. Bartók's message was simple: 'Liberation from German ideological, economic, cultural influences'.[72] He was, therefore, disbelieving of the *naïveté* of the British Prime Minister, Neville Chamberlain, in reaching the Munich Agreement in September 1938.[73] Bartók saw that this just worsened the situation. A great settling of accounts had to come, and the later it came the more frightful it would be:

Meantime, so much trouble has broken out all over the world, such unrest, such upheavals—and now this shocking change of front on the part of the Western countries. One ought to get away from here, from the neighbourhood of that pestilential country, far, far away, but where: to Greenland, Cape Colony, the Tierra del Fuego, the Fiji Islands, or somewhere even the Almighty has not heard of![74]

---

[69] (London, 1958), 264–5. According to Ove Nordwall the work was Mozart's Concerto for two pianos, K 365 ('Béla Bartók and Modern Music', *Studia Musicologica*, 9 (1967), 275).
[70] *BBL* p. 268.   [71] Ibid. 270–1.   [72] Ibid. 270.   [73] Ibid. 271–3.
[74] Letter, Bartók to Mrs Zoltán Székely, 24 Oct. 1938, in various languages (this section in Hungarian), reproduced in English in *BBL* pp. 273–4.

Part II

TWO RELATIONSHIPS

7·
BARTÓK, HESELTINE, AND GRAY

Among Bartók's early supporters in Britain few demonstrated their enthusiasm for his music as keenly and personally as Philip Heseltine (alias Peter Warlock) and Cecil Gray. At the height of their involvement with Bartók during the early 1920s both were in their twenties and attempting to forge careers as writers about music, but devoting much time to composition as well. (Posterity has recognized the greater composer in Heseltine, the better critic in Gray.) Not a little of the new level of recognition accorded to Bartók in Britain after the First World War can be attributed to their essays and reviews. Indeed, so effective were these early writings in evaluating Bartók's output and significance that they fathered a critical tradition, some elements of which still linger in the Bartók literature of today.

Heseltine was introduced to the music of Bartók during the years 1908–11, while he was still a schoolboy at Eton. His enlightened music master was Colin Taylor, who later wrote of their lessons together:

Sensing, and I hope rightly, that had I insisted on the stereotyped drill commonly meted out to those in my charge, the boy as likely as not would give up music altogether, the upshot was that I devoted the greater part of lesson time to an attempt to enlarge his musical horizons. . . . In those days I probably considered myself the deuce of a go-ahead modern, for I was playing and teaching Debussy, Ravel, Schoenberg, Scriabin and the then available Bartók.[1]

Yet from his Eton days Heseltine imbibed more immediate interest in the music of Delius and Schoenberg than of Bartók. During the pre-war years he collected, studied, and transcribed a vast quantity of Delius's music,[2] and in 1912, as the result of some months spent in Germany, he produced one of the earliest articles on Schoenberg's music in the English language.[3] By the time of his year in Oxford, 1913–14, Heseltine's enthusiasm for Schoenberg had ebbed—he was perplexed by the latest atonal pieces—and he began to incline more to the simplicity of Grieg's folksong settings. The retreat from Schoenberg was not, however, translated into a greater zeal for Bartók. To Colin Taylor he wrote in early 1914:

Béla Bartók, about whom my tame Parsee waxed so enthusiastic, disappointed me greatly: I got some piano pieces of his the other day, and I find them crude and

[1] 'Peter Warlock at Eton', *Composer*, no. 14 (Autumn 1964), 9. Taylor probably refers to *For Children* (1908–9) and *Ten Easy Pieces* (1908).
[2] See Ian A. Copley, 'Warlock and Delius—A Catalogue', *ML* 49 (1968), 213–18.
[3] 'Arnold Schoenberg', *MS* 21 Sept. 1912, pp. 176–8.

barbaric, with a considerable element of Hungarian folk-song and 'snappy' rhythms, but harmonically they are merely dull—some of them are in two parts throughout, but they sound strange on account of the extraordinary intervals by which the parts progress. His style is as unlike Schönberg as it is possible to imagine.[4]

Notwithstanding this disappointment, during 1914 Heseltine probably attended two London concerts featuring Bartók's works: Franz Liebich's concert of 11 March, and Henry Wood's première of the *Suite* No. 1, Op. 3 on 1 September.[5]

In the late spring of 1916 Heseltine met Gray at the Café Royal in London and a lifelong bond was formed between the two. They were immediately united in an enthusiasm for Delius's works, soon followed by a joint admiration for the music of Bernard van Dieren, a Dutch composer who had settled in Britain some years before the war. As with so many of their pursuits together, this admiration was a very active one. In February 1917 Gray used his private income to mount a concert of van Dieren's works at the Wigmore Hall. The result was disastrous: the critics reacted with complete hostility, and the £5 of receipts went little way towards covering the expenses of £110.[6] Heseltine was not to be deterred, conceiving the idea soon afterwards of a concert devoted to Bartók's works followed by one of Schoenberg's.[7] When he informed the leading critic Edwin Evans of these intentions over drinks at the Café Royal, Heseltine precipitated a riot. As Gray later recalled, Evans was so shocked at the idea of performing the proscribed works of an enemy alien that he assaulted Heseltine and threatened to break up any such performance with one of the contingents of drunken Australian soldiers then plentiful in Britain.[8] Needless to say, these concerts did not materialize at this time, although the idea was resurrected after the war.

During 1917 Heseltine's interest in Bartók's music grew. In contrast with his pre-war views, he now saw Bartók as a leading composer, similar in stature to van Dieren, Delius, and, with some qualification, Percy Grainger and Irving Berlin.[9] He elaborated this opinion in the March issue of the *Palatine Review*, replying to the critics of the recent Wigmore Hall concert with an enumeration of the virtues of van Dieren and Bartók and an attack upon the modern 'idols' of Stravinsky, Richard Strauss, and Skriabin.[10] In the summer of 1917, while staying near D. H. Lawrence in Cornwall, Heseltine tried to apply his knowledge of some of Bartók's techniques more

---

[4] Unpublished letter, Heseltine to Colin Taylor, 1 Feb. 1914, in English, BL Add. MS 54197. Heseltine is probably talking about the *Bagatelles*, Op. 6.
[5] See *MT* 63 (1922), 164. See Chapter 2, above, for details of these concerts.
[6] Cecil Gray, *Peter Warlock: A Memoir of Philip Heseltine* (London, 1934), 141.
[7] Cecil Gray, *Musical Chairs* (London, 1948), 112–14.
[8] Ibid. 112–13.
[9] See BL Add. MS 54197 (18 Jan. 1917).
[10] Untitled article, *Palatine Review*, no. 5 (Mar. 1917), 25–7.

practically. Writing to Colin Taylor he told of his attempts to set Celtic folk-tunes in a straightforward way, but employing a wider range of harmonies than was current British practice.[11] When later in the year Heseltine felt the need to move to Ireland to avoid any reassessment of his capacity for military service, he continued his experiments with folksong. But his attempts pleased him little, as he confessed to Gray in April 1918:

As far as I can see at present, it is unsatisfactory to use more than fragmentary quotations from them [folksongs] in a composition: they do not seem suitable as 'themes' for 'treatment'—they are somehow too proud as well as too perfect, complete, rounded, etc. for that. But on the other hand to make little works which structurally coincide with the structure of the melody—extending nothing, curtailing nothing—formally analogous in method to Grieg with his Norse tunes and Bartók with the Hungarian—seems foredoomed to failure.[12]

Because of the war, of course, Heseltine was not aware of Bartók's more wide-ranging uses of folk-materials in his recent compositions. Despite harbouring reservations about Bartók's methods, Heseltine featured some of his works in a lecture-recital presented at the Abbey Theatre, Dublin on 12 May 1918. This presentation was part of a decidedly eccentric series of Sunday evening addresses on such varied topics as 'How National Taste is Made' and 'The Catholic Theory of the State'. Heseltine's talk, simply entitled 'What Music Is', sought to outline the theory and practice of music over the last millennium 'from the appreciator's point of view—taking the appreciator into one's confidence, as it were, over the heads of the hierophants'.[13] To illustrate his views Heseltine drew on Irish and Indian folksongs, works by Mussorgsky, Delius, van Dieren, Chopin, Skriabin, and about six short pieces by Bartók.[14] About the evening Heseltine wrote to Gray that nearly 400 people had listened with respect but, as various questions had shown, very little understanding.[15] Although this response disappointed Heseltine, he considered that he had gained a reward, while preparing for the concert, in a heightened esteem for Bartók's *Bagatelles*, Op. 6:

The Bartók Bagatelles are magnificent: they are by far his best work. Looking at them anew after nearly a year without music I find most wonderful revelations in

[11] See Ian A. Copley, *The Music of Peter Warlock* (London, 1979), 227.

[12] Letter, Heseltine to Gray, 7 Apr. 1918, in English, reproduced in Gray, *Peter Warlock*, p. 184.

[13] Unpublished letter, Heseltine to Colin Taylor, 25 Apr. 1918, in English, BL Add. MS 54197. The first three pages of Heseltine's talk are preserved in BL Add. MS 57967. The talk began: 'I do not want to say much about the Letter of Music—about the technique of music: most of you would be profoundly bored if I did, and not in the least edified. So I shall confine myself almost entirely to the Spirit of Music—for that, after all, is What Music Is.'

[14] The event is described in Gray, *Peter Warlock*, pp. 159–60. Pieces from *Bagatelles*, Op. 6, *Ten Easy Pieces* (1908), and *For Children* (1908–9) were included in this selection.

[15] Ibid. 187.

them. . . . Last year I never hoped to be able to perform the Bartók Bagatelles, and even thought of having them cut for pianola—I am now well on the way to their mastery—even of No. X—the best (save perhaps for XII—with the repeated notes and curious little scales—which is a marvel) and the hardest of them all.[16]

At this time Heseltine was not inclined to bestow such praise on many other composers. Writing to Delius just after the Dublin engagement, Heseltine expressed the concern:

At present there is only Bernard van Dieren who can even share the name of composer with you; Béla Bartók has done some very fine small works but I have seen nothing of his that is less than ten years old. I hope the war has left him unscathed; he might become a very great man. But who else is there? Schönberg still shows a cold, white light, but he will never escape from the toils of his self-imposed originality—I cannot think of a single other name that does not seem to belong to a barrel-organ grinder or his performing monkey rather than to a composer.[17]

Encouraged none the less by the experience of his lecture in Dublin, Heseltine ventured to present further general papers on the nature of music,[18] addressing the Musical Association in London on 13 May 1919 on the subject 'The Modern Spirit in Music'.[19] Bartók was not specifically mentioned in the paper, although it revealed much about the reasons behind Heseltine's preferences in recent music and his concern at several current trends in composition. He warned that the writing of a greater variety of chords by composers had to be associated with a broadening of listeners' aural appreciation if the expressive power of music was not to be diminished. The current fascination of composers with puppets, marionettes, and other 'counterfeit human beings' was attacked as a decadent trend, which exalted the sign above the thing signified and caused spiritual realities to be ignored through a preference for mere external representations. In a following letter to the new journal *Music and Letters*, Heseltine summarized the quality for which he searched in contemporary music:

We live in an age of sensationalism, superlatives and exaggeration and it is perhaps inevitable that drums should be beaten and trumpets blown on behalf of music as of everything else: but some of us are listening more intently for the still small voice of genius who will 'take homely and familiar things and make them fresh and beautiful' at his touch.[20]

Late in 1920 Bartók's position in a trinity of leading composers was publicly asserted by Heseltine and Gray. As revealed in a letter to Gray,

[16] Unpublished letter, Heseltine to Colin Taylor, 14 May 1918, in English, BL Add. MS 54197.
[17] Letter, Heseltine to Frederick Delius, 15 May 1918, in English, reproduced in Gray, *Peter Warlock*, p. 175.
[18] See BL Add. MS 57795 (14 May 1918).
[19] *Proceedings of the Musical Association*, 45 (1918–19), 113–30.
[20] *ML* 1 (1920), 164. See also Gray, *Peter Warlock*, p. 24.

written while he was on a trip to France to visit Delius, Heseltine came to
know of several recent Bartók works, including the Second String Quartet,
Op. 17, the ballet *The Wooden Prince*, Op. 13 ('a "pantomime"—puppets,
I'm afraid, or at least puppetism'), and the *Allegro barbaro* (1911), and
immediately sent for their scores.[21] This discovery, coupled with Sir Henry
Wood's revival of Bartók's *Suite* No. 1 at the end of August, inspired
Heseltine to resuscitate his wartime plans for a concert of Bartók's
compositions. The September issue of the *Sackbut*, of which Heseltine had
recently been appointed editor, hailed Bartók as the equal of van Dieren,
second only to Delius, and gave notice of a forthcoming performance of the
First String Quartet, Op. 7.[22] By the following month Heseltine's plans had
grown. A full Bartók programme including first British performances of
both string quartets was scheduled for 8 December at Mortimer Hall, with
Heseltine and Gray as its official promoters.[23] Heseltine also revived his
idea of performing Schoenberg's works, announcing at the same time the
British première of *Pierrot lunaire*, Op. 21 on 25 January 1921. As in 1917,
however, these plans were not to be realized. Financial problems began to
plague the *Sackbut*, resulting in a change of ownership and a reassessment
of the journal's commitments. Gray, perhaps recalling the fiasco of the van
Dieren concert, had for some time harboured doubts about the viability of
this ambitious series. Heseltine reluctantly came to accept his point of
view.[24] The fate of these potentially historic performances was formally
announced in the journal's November issue with the curt sentence: 'The
Sackbut concerts are, for the present, suspended.'[25]

By early November Heseltine thought it was time to establish personal
contact with Bartók. In a letter, now lost, he expressed his admiration of
Bartók's music, told of the *Sackbut*'s plans to promote it, and outlined his
disappointment in the works of the current favourites, Debussy and
Stravinsky.[26] Copies of the first six issues of the journal were sent along
with the letter. Bartók was overjoyed at this British statement of support,
both unsolicited and unexpected. Only in recent months, with the calming
of the political situation in Hungary, had he been able to turn his mind to re-
establishing contacts abroad, and Heseltine, as editor of a London journal,
would certainly be a most valuable ally. Bartók's (first) wife, Márta,
excitedly wrote with a full description of Heseltine's letter to Bartók's
mother in Pozsony, proudly telling of the journal's placement of Bartók
among the greatest living composers and mentioning the possibility of a visit
to England.[27] A few days later Bartók himself wrote to Heseltine a long,

[21] BL Add. MS 57794 (21 Aug. 1920). See also BL Add. MS 54197 (17 Oct. 1921).
[22] *Sackbut*, I (1920–1), 220–1.   [23] Ibid. 279.
[24] See BL Add. MS 57794 (24 Oct. 1920).
[25] *Sackbut*, I (1920–1), 322.
[26] Deduced from VBBH pp. 139–40, and BBcl pp. 308–10.
[27] BBcl pp. 308–10.

enthusiastic letter of reply, quite out of keeping with his customary reserve in correspondence.[28] He listed various of his works which were still unknown to Heseltine.[29] He further acknowledged the peasant music of Hungary and neighbouring regions as the main influence on his composition, and drew Heseltine's attention to the work of his compatriots Zoltán Kodály and László Lajtha. At Heseltine's attacks on Stravinsky and Debussy, however, he could only express surprise. While painstakingly writing his reply (beginning in English but soon retreating into French), Bartók received a second letter from Heseltine which asked him to contribute occasionally to the *Sackbut*. His agreement led to the article 'The Relation of Folk Song to the Development of the Art Music of our Time', which appeared in the journal's issue for June 1921.[30]

Meanwhile, in mid-November 1920 Gray's controversial article, 'Béla Bartók', had appeared in the *Sackbut*.[31] Writing in 1965 Denijs Dille called on this article as evidence that Gray was the most informed Bartók critic of the early 1920s, except for Kodály, and was even led incorrectly to suggest that it was Gray who must have drawn Heseltine's attention to Bartók's music.[32] Bartók himself was greatly impressed by the standard of the writing, especially as it was by someone quite unknown to him. Early in the New Year he wrote to Heseltine (this time in German) with a detailed assessment of the article, mainly of praise, but including the advice that the influence of folk-music upon his works had been understated.[33] Bartók concluded with a confession that having just received some of Stravinsky's recent compositions he had come to see Heseltine's point in attacking them.

Soon Gray also started a correspondence with Bartók. In the first extant letter of this exchange, dated 16 January 1921, Bartók mentioned his tentative plans to visit London, and warmly invited Gray to stay with his family in Budapest.[34] Bartók was also appreciative of Gray's influence in securing an invitation from the *Chesterian* for an article, resulting in 'The Development of Art Music in Hungary' in the January 1922 issue of that journal.[35] As 1921 progressed the friendship between Bartók and his two English correspondents quickly deepened. Bartók's letter of 7 February to Heseltine, written in English, shows just how useful the Hungarian realized this new relationship could be for the advancement of his music.[36] Because

---

[28] Letter, Bartók to Heseltine, 24 Nov. 1920, VBBH pp. 139–40. The tone of the letter is discussed in Denijs Dille, 'Vier unbekannte Briefe von Béla Bartók', *Oesterreichische Musikzeitschrift*, 20 (1965), 456.   [29] Listed in *Sackbut*, 1 (1920–1), 374.
[30] *Sackbut*, 2 (1921–2), 5–11.
[31] The contents of Gray's essay are outlined in Chapter 2, above.
[32] Dille, 'Vier unbekannte Briefe', p. 459.
[33] Unpublished letter, Bartók to Heseltine, 8 Jan. 1921, SzFAC.I. See also *BBcl* pp. 310–13.
[34] See *BBlev*.v pp. 263–4 and BL Add. MS 57784.
[35] *Chesterian*, NS no. 20 (Jan. 1922), 101–7. This essay is discussed in Chapter 2, above.
[36] Letter, Bartók to Heseltine, 7 Feb. 1921, VBBH pp. 140–1.

he had not heard from Heseltine for some time, Bartók listed the various items that he had recently sent to him: scores of four compositions, an article for the *Sackbut*, his long (but unanswered) letter of the previous month, and, through his publishers, two further scores. The letter also contained some informal assessments of recent work by Malipiero, Goossens, and Pizzetti, and again mentioned Bartók's disappointment at some of Stravinsky's recent pieces. But Bartók was still mystified by the references of both Gray and Heseltine to van Dieren: 'Now all my hope is in van Dieren! How can I get his works? What did he compose? I made [*sic*] the same questions to Mr. Gray in a recent letter, but—it seems—he had forgotten to inform me about this matter in answering my letter.'

In the journals Heseltine and Gray maintained their promotion of Bartók's music. The March issue of the Viennese periodical *Musikblätter des Anbruch*, a celebratory number in honour of Bartók's fortieth birthday, contained an abridged German translation of Gray's *Sackbut* article.[37] Using the pseudonym Barbara C. Larent, Heseltine penned a review in the *Sackbut* entitled 'Sir Charles Snarls', in which he attacked Sir Charles Stanford for presenting a paper on tendencies in recent composition which failed to take account of the work of such composers as Schoenberg, Stravinsky, and Bartók.[38] Under his own name Heseltine also wrote a glowing appreciation of Bartók's *Three Studies*, Op. 18 of 1918, which history has not backed up. His comments are interesting, however, as they mention certain reservations about earlier Bartók works which Heseltine had not previously voiced in public:

These pieces [Op. 18] are not merely the best Bartók has yet written for the pianoforte. They are a contribution to piano literature to rank alongside of the *Etudes* of Chopin and Liszt. . . . Many of his earlier piano pieces, without being exactly experimental yet seemed somewhat fragmentary, like sketches made in preparation for some longer and more comprehensive work. These *Etudes* are more extended in form and although, true to their designation, they offer plenty of novel problems of technique to the ambitious pianist and provide much that will interest the seeker of new and unfamiliar effects of piano timbre and colour, technique is never for a moment made subordinate to the expressive purpose. Bartók is a consummate technician with a fully-developed style, that is all his own, at his fingers' ends; but the technical interest of this work is incidental for—quite apart from technique—it is great music.[39]

The exchange of letters continued, with Heseltine writing to Bartók in late February or early March 1921 about his interests in old English music and the work of Delius. In reply on 17 March Bartók undertook to try to

---

[37] 'Béla Bartók', *Musikblätter des Anbruch*, 3 (1920–1), 90 ff.
[38] *Sackbut*, 1 (1920–1), 419. Stanford's paper is reproduced in the *Proceedings of the Musical Association*, 47 (1920–1), 39–53.   [39] *Sackbut*, 1 (1920–1), 426.

arrange a performance of Delius's Violin Concerto (1916) in Budapest, and sent a vocal score of his opera *Duke Bluebeard's Castle*.[40]

Later in March, oppressed by the continuing financial difficulties of the *Sackbut*, Heseltine left Britain for a period of foreign travel, in the company of Gerald Cooper, a budding London concert entrepreneur. After a short visit to North Africa Heseltine travelled to Budapest, arriving on about 7 April, and stayed with the Bartóks.[41] Many years later, Bartók's first wife's strongest recollection about this visit concerned Heseltine's personal appearance: pink shirt, lilac-coloured cravat, and reddish beard![42] Bartók and Heseltine appear to have established a warm relationship, as is seen in a letter written by Heseltine while he was in Budapest to Delius: 'Bartók is quite one of the most lovable personalities I ever met.'[43] To his mother Heseltine wrote about material conditions in Hungary:

We have had a lovely time here in Budapest and everyone has been most kind and hospitable to us. The country is slowly recovering from the effects of the war, but it is quite terrible to hear—and indeed to see—how people like the Bartóks and Kodálys have suffered, from lack of food and fire, these last years—and here in Hungary political troubles have made life even more difficult than it need have been.[44]

The following month Gray, too, visited Bartók in Budapest, although he lodged with the Kodálys.[45] In his autobiography *Musical Chairs*, Gray recalled the visit: the poverty in Hungary, the friendliness of the Hungarian composers, and Bartók's complete absorption in his musical tasks.[46] But during this visit Gray came for the first time to realize the trend of Bartók's most recent compositions towards the unrelieved dissonance of atonality, a far cry from the tonal charm of his pre-war folksong settings:

I remember only too well the occasion when I was first introduced to the recently completed music of the ballet, *The Marvellous Mandarin*. It was a day of scorching heat in August [*sic*] (in Budapest), in a street in the centre of the town where the din of the traffic was intolerable; and Bartók played to me the piano score on an inferior instrument which was also decidedly out of tune. Even in orchestral performance this work is a severe strain on the nervous system, to say nothing of one's aesthetic sensibilities; consequently my sufferings under the aggravated circumstances of the

[40] VBBH p. 141.   [41] Gray, *Peter Warlock*, p. 213.
[42] Dille, 'Vier unbekannte Briefe', p. 456. In conversation with the author on 10 Feb. 1984 Béla Bartók, jun. also mentioned Heseltine's singular appearance.
[43] Letter, Heseltine to Frederick Delius, n.d. [mid-Apr. 1921], in English, reproduced in Gray, *Peter Warlock*, p. 213.
[44] Unpublished letter, Heseltine to his mother (Mrs Buckley Jones), 20 Apr. 1921, in English, BL Add. MS 57961. Because of their membership of the Music Directorate of the left-wing Kun regime of 1919, Bartók and Kodály had been ostracized by the succeeding right-wing government of Miklós Horthy and had only retained their positions at the Budapest Academy of Music with difficulty.
[45] See BBlev.v p. 267.   [46] *Musical Chairs*, pp. 180–5.

performance can perhaps be estimated, without much exercise of imagination. I have seldom suffered so much from music. It was even too much for the composer himself. About three-quarters' way through the work he suddenly stopped playing, saying that he did not feel in the right mood. I searched my brains desperately for some complimentary and appreciative words which would not sound too insincere, but failed. I could find nothing to say.[47]

Heseltine had also become anxious about Bartók's interest in ballet. Describing his visit to Budapest in a letter to Colin Taylor, he stated: 'Bartók has done some stupendously great works—great in conception and idea, not in size—always *multum in parvo*. But he has, alas, now got the ballet craze—rather late in the day—which is a pity.'[48] Measured by the comments about their times in Budapest and subsequent correspondence, Heseltine appears to have established a stronger personal bond with Bartók, while Gray was more drawn to his host, Kodály.[49]

In Britain, meanwhile, Heseltine and Gray had lost the major vehicle for expressing their views with the enforced change in staff of the *Sackbut* after its June 1921 issue,[50] but their desire to promote Bartók's cause was undiminished. In a postcard of 17 August Heseltine suggested to Gray that he might undertake a translation of the libretto of *Duke Bluebeard's Castle* and also write an article for the *Daily Telegraph* on 'A Visit to Budapest'.[51] Gray did not immediately respond to these ideas, but later in the year Heseltine himself submitted a short account of the work of Bartók and Kodály to the *Daily Telegraph*. It was returned so quickly that he felt sure it had not even been read.[52]

By mid-1921 the renewal of contact with the critic M.-D. Calvocoressi had encouraged Bartók to intensify his planning for a British concert tour. In his reply to Calvocoressi's initial letter pledging support to his cause, Bartók mentioned that his two London friends, Heseltine and Gray, would also be pleased to see his concert plans reach fruition, although he doubted if they had the ability to organize such a tour themselves.[53] In this judgement Bartók was perfectly realistic. He could imagine that, in spite of their enthusiasm, the youthful Heseltine and Gray exerted little influence

---

[47] Ibid. 182. According to Béla Bartók, jun., Gray stayed in Budapest from 20 May to 1 June (*Apám életének krónikája* (Budapest, 1981), 187). It is interesting to note that in other correspondence at this time Bartók expressed the opinion that *The Miraculous Mandarin*, Op. 19 was his best composition. See BBrCal pp. 202–5.

[48] Unpublished letter, Heseltine to Colin Taylor, 17 Oct. 1921, BL Add. MS 54197.

[49] Thirteen letters from the Kodálys to Gray are held in BL Add. MS 57785. They date from 26 Apr. 1921 to 12 Nov. 1948.

[50] The extent of the *Sackbut*'s losses was considerable. Between Nov. 1920 and Mar. 1921 receipts totalled £130 and expenses £352. See BL Add. MS 57794 (21 Oct. 1921).

[51] BL Add. MS 57794 (17 Aug. 1921).

[52] BL Add. MS 57794 (6 Dec. 1921).

[53] BBrCal pp. 202–5. The significance of Calvocoressi's offer to help promote Bartók's music in Britain is discussed in Chapter 2, above.

over London's musical life, especially since the *Sackbut* had been removed from their control. Although Heseltine and Gray did still try to arrange a tour, it was, as Bartók had expected, Calvocoressi's interventions which achieved results, albeit slowly. Writing to Gray in January 1922, two months before Bartók's tour took place, Heseltine noted with bitterness:

I also heard from Bartók. It is very significant that d'Arányi [Jelly Arányi] can be moved to activity on his behalf by Calvocoressi, a respectable and influential critic of several years' standing, whereas by *our* entreaties she is only moved to expressions of admiration unsupported by any activity whatsoever.[54]

While these concert plans were crystallizing, Bartók's correspondence with Heseltine and Gray continued spasmodically. In a letter to Gray written on 15 October he was mildly reproving, mentioning that he had not received a copy of his June *Sackbut* article, nor heard anything from Heseltine since his departure from Budapest.[55] But elsewhere in this rather long letter he wrote openly about his concert plans and bemoaned the constant deterioration in Hungary's economic and musical circumstances. By calling Gray's attention to his *Sackbut* article and the lack of news from Heseltine, Bartók was probably hoping for a very practical response: payment. And he was successful. The following month, with some sense of guilt, Heseltine arranged for copies of the *Sackbut* and a bank draft for nearly £5 to be sent to 'the unfortunate Bartók'.[56] On 29 December Bartók wrote more cheerful letters to both Heseltine and Gray (in English and French respectively), urging them to attend the Frankfurt-am-Main performance of his stage works, then planned for April, and giving notice of his most recent composition, the (First) Violin Sonata, which he hoped to play with Jelly Arányi if the much-discussed London concerts ever materialized.[57] Already thinking of accommodation costs, Bartók asked Gray if he might stay with him in London for about a fortnight. (By this time Heseltine had withdrawn to his family home in Wales.) Whatever Gray's answer, Bartók eventually stayed for most of his visit at the South Kensington home of Duncan and Freda Wilson—an arrangement made by the Arányis.[58]

In early 1922 Heseltine and Gray continued to act as British watchdogs of Bartók's reputation. On 6 January, for instance, Heseltine suggested to Gray that he might attend a coming lecture of the Musical Association entitled 'Modern Developments in Music', to be delivered by Eugene Goossens.[59] At a previous talk Goossens's 'facts were as muddled as his opinions and poor Bartók was put down as a disciple of Schönberg. In view

[54] Unpublished letter, Heseltine to Gray, 6 Jan. 1922, BL Add. MS 57794.
[55] *BBlev.*v p. 269.   [56] BL Add. MS 57794 (19 Nov. 1921).
[57] VBBH p. 141, and *BBlev.*v p. 270, respectively.
[58] See Joseph Macleod, *The Sisters d'Arányi* (London, 1969), 138–40.
[59] *Proceedings of the Musical Association*, 48 (1921–2), 57–76.

of Bartók's impending visit, it would be as well if somebody were present to contradict this blasphemy if it should be repeated.'[60] Goossens did not commit such 'blasphemy', however; in fact he refrained from mentioning Bartók's name altogether. The next month, in their eagerness to promote modern Hungarian music, Heseltine and Gray found themselves cast in competitive roles by the editor of the *Musical Times*, Harvey Grace.[61] Grace had accepted Heseltine's short article, earlier rejected by the *Daily Telegraph*, and it duly appeared in the journal's March issue, along with a short review by Heseltine of two recent compositions of Kodály.[62] But when Gray submitted an article on Kodály to the same journal, probably during January, Grace showed no interest. This prompted Heseltine to write apologetically to Gray for having obstructed his article, and to promise to remonstrate with Grace.[63] The matter was eventually resolved by Grace's commitment to publish Gray's article in the May issue.[64]

Despite the considerable promotional activity, there was no certain way in which a conservative public could be cajoled into a hasty acceptance of Bartók's merits. The outcome of his first visit, in March 1922, could not be forecast. In a letter to Gray dated 12 March, Kodály wrote that he would be happy if Bartók's visit did nothing more than restore his goodwill and humour.[65] But the tour was decidedly a success, with a comprehensive media coverage and a surprisingly open-minded approach from many reviewers. For Gray, none the less, the performances of the First Violin Sonata (1921) reinforced his apprehension of the dissonant direction of Bartók's composition. While he maintained strong public support for Bartók, his private thoughts were somewhat different, as he confessed in a personal note written on 23 March:

Bartók by the way is here just now and gives a concert tomorrow. I am reminded of him by speaking of Beethoven. I recollect with a certain shame that I speak of him in terms one should only reserve for the very greatest which Bartók is not by a long chalk. But it is attributable to the enthusiasm of finding a man with a few at least of the qualities of Beethoven—and after all crimes committed in the name of enthusiasm are easily forgiven.[66]

Heseltine appears to have maintained a greater genuine enthusiasm and stronger personal relations. He heard Bartók perform in London, was only prevented by train difficulties from joining him in Aberystwyth, and even

[60] Unpublished letter, Heseltine to Gray, 6 Jan. 1922, BL Add. MS 57794.
[61] BL Add. MS 57794 (8 Feb. 1922).
[62] BL Add. MS 57794 (23 Dec. 1921); 'Modern Hungarian Composers', *MT* 63 (1922), 164–7; review, ibid. 187.   [63] BL Add. MS 57794 (8 Feb. 1922).
[64] 'Zoltán Kodály', *MT* 63 (1922), 312–15.
[65] BL Add. MS 57785 (12 Mar. 1922).
[66] Personal note, dated 23 Mar. 1922, under 'Views and Reviews' section of Cecil Gray Papers, BL Add. MS 57788.

entertained Bartók at his Welsh home in Abermule, near Newtown.[67] In a letter to Gray from Abermule on 25 March, Heseltine wrote excitedly about Bartók's 'marvellous' *Improvisations*, Op. 20, finding the 'least good' of the series the seventh piece, which had been written 'à la mémoire de Claude Debussy'.[68] He also expressed a desire to go to Frankfurt-am-Main to hear Bartók's stage works if suitable sponsorship could be arranged. This was quite a change from the sentiments of a letter of 6 January in which he had written to Gray:

I don't feel inclined to go to Frankfurt—in fact I am becoming more hermit-like every day and have little inclination even to go as far as London, unless compelled by necessity. And although 'Bluebeard' is undoubtedly a fine work, it doesn't seem to me to contain anything essential of Bartók which has not manifested to even greater advantage in the quartets or piano music.[69]

Knowing of Heseltine's new interest in the Frankfurt performances, Bartók sent him two postcards in April to make sure that he was aware of their postponement until May.[70] It appears that ultimately Heseltine did not manage to attend.

After this 1922 tour, Bartók's relationship with Heseltine and Gray was not maintained at its previous level. As far as can be ascertained there was a drifting apart in interests rather than any confrontation. Heseltine, in his Welsh isolation, was entering a most prolific phase as a composer, shedding his earlier cosmopolitan enthusiasms as he developed a more determined, unified character. If Gray's personality characterizations are accepted,[71] Bartók was a passion of the critic Heseltine rather than of the composer Warlock. For Gray, too, the friendship with Bartók was becoming increasingly uncomfortable because of his inability to accept Bartók's more radical musical expression. An undated jotting among Gray's papers most succinctly states his perception of the developments of Bartók's middle period: 'The greater the freedom in tonality the more the restrictions in other ways . . . Bartók, freest of spirits, became his own gaoler.'[72] Over the coming years Gray came more and more to elaborate on this position in his public writings. From Bartók's perspective, the relationship had been an unexpected boon, aiding enormously the promotion of his music and the establishment of an international reputation. Perhaps to the annoyance of

[67] See, respectively: BL Add. MS 57794 (25 Mar. 1922); MTA BA-B 3477186; Ian Parrott, 'Warlock and the Fourth', *MR* 27 (1966), 130.

[68] BL Add. MS 57794 (25 Mar. 1922).

[69] Unpublished letter, Heseltine to Gray, 6 Jan. 1922, BL Add. MS 57794.

[70] Unpublished postcards, Bartók to Heseltine, 11 Apr. and 30 Apr. 1922, in English, in Mary Flagler Cary Music Collection of Pierpont Morgan Library (New York).

[71] Cecil Gray, 'Peter Warlock (Philip Heseltine)', *Chesterian*, NS no. 40 (June 1924), 245–50.

[72] Personal note, n.d., under 'Views and Reviews' section of Cecil Gray Papers, BL Add. MS 57788.

Heseltine and Gray it had given Bartók the confidence to foster other British contacts, with such musicians as Calvocoressi, Dent, and Hull, which he realized would be of greater assistance in furthering his interests during subsequent years.

The decline of the relationship and the various activities of its three participants are interesting to document, particularly in light of the criticisms levelled by Gray at Bartók's post-war music. On the Hungarian's visit to London in May 1923 the three met at his Contemporary Music Centre concert, which included performances of both violin sonatas. In a letter to his first wife, Bartók wrote of the great enthusiasm of Heseltine and Gray at this event.[73] Gray's review of the concert for the *Music Bulletin* bore no hint of adverse opinion. He asserted Bartók's affinity with Beethoven, lauded his ability perpetually to renew his compositional technique, and explained the 'novel and wholly logical' concept behind these works, namely, the opposition rather than blending of the two parts.[74] In conclusion, Gray praised the superb playing of Jelly Arányi; Bartók's piano-playing was nowhere discussed. Favourable mention of Bartók's work was also found during 1923 in Heseltine's book *Frederick Delius*. Heseltine cited Delius's operas *A Village Romeo and Juliet* (1900–1) and *Fennimore and Gerda* (1909–10) as the only contemporary stage works, besides Bartók's *Duke Bluebeard's Castle* (1911), which really tried to strike a balance between real dramatis personae and mere personifications of passions.[75]

Although Bartók was so encouraged by his reception in Britain that he ventured a second tour later in 1923, there is no evidence of any contact between the three on that occasion. Certainly, Bartók's schedule would have afforded little time for socializing.[76] The following year Gray's book *A Survey of Contemporary Music* was published, in which a chapter was devoted to Bartók.[77] This chapter was a composite product, based on Gray's 1920 *Sackbut* article, but also incorporating much of his review of the concert of 7 May 1923, and small, incongruous additions of later date. Most important of these additions was the final paragraph, where Gray aired some of his anxieties, until then only privately expressed. He wrote in part:

> There is a definite conflict in his [Bartók's] art between his simple, profound, and Beethovenian nature and the nervous exasperation and feverish restlessness which are so typical of the modern spirit. . . . Greatly though one admires the second string quartet and the violin sonatas, one cannot help noticing in them a certain lack of

[73] *BBcl* p. 340.
[74] 'Contemporary Music Centre Concert, May 7, 1923', *MB* 5 (1923), 191–2.
[75] *Frederick Delius* (London, 1923), 79. For earlier statements of this ideal by Heseltine see: Gray, *Peter Warlock*, pp. 131–5; Sir Thomas Beecham, *Frederick Delius* (London, 1959), 178.
[76] See Bartók, jun., *Apám életének krónikája*, pp. 212–13.
[77] (London, 1924), 194–209.

harmonious balance and equipoise—an almost frightening brutality and explosive violence which do not compare favourably with the serenity and self-mastery of the first quartet, which is perhaps the highest point to which Bartók has ever attained. Together with this one observes a tendency towards a hardening and formularization of his harmonic idiom. . . . His fondness for dissonance of the most aggravated type, as exemplified particularly in the second violin sonata, seems to be in danger of mastering him altogether, and becoming an obsession, as it has with Schönberg.[78]

During 1924 and 1925 Heseltine and Gray probably had no contact with Bartók, until the Hungarian sent Gray a miniature score of his popular *Dance Suite* (1923) for Christmas 1925.[79] Gray replied with a short, polite letter, saying how good it was to hear from Bartók 'after such a long time'.[80] He went on to mention various recent British performances of Bartók's works, and to express the hope that Bartók would visit Britain again soon. A silence of many years followed.

One of Heseltine's last written references to Bartók, before his untimely death in 1930, can be found in a letter to Colin Taylor of 19 January 1929.[81] There, Heseltine wrote of his current dissatisfaction with Delius's music—its lack of formal construction, thickness of texture, and sweetness of harmony—and how such music made him long for the precision of the Elizabethans or Mozart, 'or else for the stimulating harshness and dissonance of Bartók, and the Stravinsky of *Le Sacre du Printemps*'. The expression of a similar opinion by Heseltine is recorded in Eric Fenby's book *Delius as I knew him*.[82] Fenby has preserved a short exchange which took place between Delius and Heseltine at Delius's home in Grez-sur-Loing. It concerned a recent radio broadcast. (Delius had himself corresponded with Bartók a considerable number of times before the First World War, and had helped him with a problem of copyright for one of his works.[83])

'What is that, Fred, that you are talking about?' asked Heseltine.
'Oh, Bartók's Fourth Quartet,' replied Delius. 'Did you hear it, Phil?'
'Yes.'
'So did I. I thought it was dreadful! I'm sick and tired to death of all this laboured writing, all this unnecessary complication, these harsh, brutal, and uncouth noises. How anybody can listen to such excruciating sounds with understanding and pleasure is beyond me! What did you think of it, Phil?'
'I'm sorry, Fred, but I don't agree with you. I think it's a masterpiece. For sheer beauty of sound it is one of the wonders of music.'
'Well,' sighed Delius, shaking his head, 'well! . . .' and relapsed into silence.

---

[78] Ibid. 209.   [79] Now held in SzFAC.I.
[80] Letter, Gray to Bartók, 28 Dec. 1925, in English, *DB*.iii p. 129.
[81] See Copley, 'Warlock and Delius', p. 214.   [82] (London, 1936), 61.
[83] See *BBL* pp. 104–12. On first meeting Delius in Zürich in 1910 Bartók felt an immediate affinity with him, and later wrote: '. . . I have never before met anyone to whom from the very first I could feel so close' (p. 104).

Gray continued to write from time to time about Bartók's music over the next twenty years. In an article entitled 'Tonality and Atonality' of 1928 he pointed out how few composers had followed Schoenberg's atonal path, and mentioned Bartók's music as 'melodically speaking at least, still fundamentally tonal in its implications'.[84] His volume *Sibelius* of 1931 expounded the theory that the music of the previous hundred years revealed a continuous expansion of all kinds of tonal resource, and he accordingly asserted that composers such as Schoenberg, Bartók, and Stravinsky were representatives of the final stage of the Romantic movement rather than of the beginning of a new epoch.[85] Some time after this volume appeared Bartók and Gray met, probably during Bartók's visit to London in November 1933. The listing of Cecil Gray as a fellow Honorary Vice-President of Chisholm's Active Society in Glasgow had prompted Bartók to seek out Gray's address.[86] Gray's only recorded recollection of the meeting concerned Bartók's amazement that a musician should possess a library as large as Gray's.[87]

This meeting with Bartók did nothing to mitigate Gray's criticism of his music. In *Predicaments* of 1936 he unfolded his most extensive and unrestrained critique of Bartók.[88] With his 'added note' obsession, Bartók had entered a harmonic cul-de-sac, Gray claimed, which was the last possible extreme of the Romantic movement: 'The final stage in the process is reached by taking a chord, adding adjacent semitones to each constituent note, taking away the harmony you first thought of—and the answer is later Bartók!'[89] To Gray, Bartók had no option but to apply a drastic simplification to his harmonic writing. He still hoped that this might come about, but saw no evidence that Bartók was yet moving in such a direction:

The influence of Stravinsky is not apparent in the most recent works, yet at the same time he does not seem to have regained his own personality, but to have become obsessed by still another alien if unidentifiable one, chiefly characterized by an insatiable appetite for unrelieved dissonance of the most exacerbated variety. His fourth string quartet is a particularly painful example of this. . . . There are definite limits to what the ear can endure, and Bartók in his later works frequently oversteps them.[90]

Despite these deficiencies in Bartók's later works, Gray still credited him with one of the most creative musical imaginations of the day, which prevented him from stagnating and gave reason to hope that he would realize his potentialities more successfully in the future.[91] A decade later, in

---

[84] *The Nation and the Athenaeum*, 5 May 1928, pp. 139–40.
[85] (London, 1931), 195–6. This theory is relevant to the compositional approach of Heseltine, Gray, and several other British composers of their day. See Arnold Whittall, 'The Isolationists', MR 27 (1966), 122–9.　　　[86] MTA BA-B 3477/67 and MTA BH 541.
[87] *Musical Chairs*, pp. 180–1.　　　[88] (London, 1936), 137, 234–5, 269–78.
[89] Ibid. 235.　　　[90] Ibid. 272–4.　　　[91] Ibid. 275.

the essay 'Contingencies' (1947), Gray briefly addressed the problem of the unpopularity of composers such as Bartók, Schoenberg, and Webern with concert audiences, concluding that much good music would always be above the head of the average man and was none the worse for that, but currently, great music that could gain popularity was the most needed commodity.[92] Berg was considered by Gray a composer of such music.

A final volume, *Musical Chairs*, consisting of Gray's memoirs, was published in 1948, three years before his death.[93] Writing so soon after Bartók's own passing, Gray devoted considerable space to a summary of their relationship and an overall evaluation of Bartók's significance as a composer. The greatest surprise is to be found in Gray's rehabilitation of Bartók to the ranks of the truly great.[94] In the mellower, final works of Bartók's American years, Gray found his hopes of a simpler harmonic style fulfilled, and those potentialities ever present in Bartók's writing at last adequately realized. The great Bartók of the First String Quartet had been regained, his stylistic extremes had been tempered, and Gray's anxieties of twenty years were allayed. Contrasting Bartók as musician and man, Gray observed how the element of sadism and laceration so prominent in the middle-period compositions found no place in Bartók's personal relations, which were always characterized by gentleness and courtesy—quite the reverse of Delius.[95] In his final portrait, however, Gray chose to stress one characteristic above all: Bartók's single-minded dedication to music.

Béla Bartók, in short, was completely inhuman. He hardly existed as a personality, but his impersonality was tremendous—he was the living incarnation and embodiment of the spirit of music. He was pure spirit, in fact, and his frail, intense and delicate physique gave the impression of something ethereal and disembodied, like a flame burning in oxygen. No need to inquire, no need to know, the cause of his death: he consumed himself, burnt himself entirely away in the fire of his genius and of his selfless devotion to his art.[96]

[92] *Contingencies and Other Essays* (London, 1947), 46.
[93] (London, 1948).   [94] Ibid. 182–3.
[95] Ibid. 187.   [96] Ibid. 181.

8.

BARTÓK AND THE ARÁNYI SISTERS

'It is good and great that I should have inspired that gorgeous sonata—but apparently a woman can't inspire the soul of a man without doing great harm. It is sad, too sad, that I should make this great man suffer.'[1] Suffer Bartók did, emotionally and professionally, at the rejection of his overtures towards Jelly Arányi. In 1922 she was in the prime of life: light-hearted, vibrant, a consummate violinist, and unmarried, although in her late twenties. Already she had excited the imaginations of many famous men, including Edward Elgar, Donald Tovey, and Aldous Huxley.[2] In this year, despite some international recognition of his talents, Bartók was an unhappy man. He was well into middle age, poor, artistically frustrated, and not altogether happy domestically. Only Jelly Arányi had in the previous year been able to dispel his depression and inspire him to write a violin sonata so promptly. Only through Jelly and her sisters had Bartók gained sufficient, firm commitments to come to Britain and score such artistic successes. But while she enjoyed playing with Bartók, matching his rhythmic energy and relentlessness, Jelly did not wish to broaden her friendship with this apparently humourless Hungarian, so devoid of the sparkle and *élan* of her social set. Through a lack of congruence in private intentions their professional collaboration was brief, barely spanning two years. These private differences, consequently, became a hindrance to the promotion of Bartók's music in Britain during the 1920s. Only with the forging of links with the BBC from the middle of the decade was the advocacy of his music seriously and systematically undertaken.

Bartók had known Jelly Arányi since her childhood. He had then been a student at the Budapest Academy of Music, she, the younger sister of one of his piano students, Hortense (Titi). Understandably, Bartók had been more involved with his pupil and her elder sister, the violinist Adrienne (Adila), than with young Jelly. None the less, he did not forget, in December 1902, to send her a card of greetings showing two young girls and their doll beside a Christmas tree.[3] From the start, Bartók enjoyed his contact with the family. Writing to his mother in late March 1902 he had told of his new teaching responsibilities:

My other pupil is Hortánsz [*sic*] Arányi, a 15-year-old girl who is the daughter of Police Superintendent Arányi. Adrienne, her elder sister (by one year), goes to the

[1] Jelly Arányi, diary entry, Apr. 1922, reproduced in Joseph Macleod, *The Sisters d'Arányi* (London, 1969), 139–40.
[2] See Macleod, *Sisters d'Arányi*, *passim*, in particular chs. 8–10.
[3] SzFAC.I (24 Dec. 1902).

Music Academy where she knows me and suggested that I teach her sister. I was almost sorry that I ask them for three forints an hour, because they live in rather humble circumstances. . . . this girl is preparing for the Academy, and so takes her work seriously. And so far she has had good teachers, so she will be little trouble for me.[4]

By November 1902 Bartók had come to know the family quite well, especially Adila. When issued with an invitation to attend a Sunday afternoon gathering at their home in Szövetség utca, however, he hesitated. While keen to meet some of the other young musicians invited, in particular the cellist Kerpely and violinist Sabathiel, he felt a little afraid of them.[5] The real draw for Bartók was the Arányis themselves:

this Arányi family is very interesting: firstly, because they are closely related to Joachim. (The Arányi girls' paternal grandmother was Joachim's sister.) Secondly, because you never hear a word of German in this family, simply because no-one in the family knows German. In this regard they are perhaps unique in Budapest. (They know French instead of German.)

An ardent nationalist, Bartók had to admire such absence of German cultural contamination. Even so, it was only on the Wednesday before the party that Bartók finally decided that he would dare to attend. To his mother he pleaded 'curiosity' in explanation of his decision:

Nevertheless I went there on Monday to find out exactly what kind of entertainment it will be. When I came to the Arányi girls—that is Titi and Adila—now no longer in the role of 'teacher', they seemed exceedingly high-spirited and they almost turned the whole house upside down. They held out the promise that the tumult will be even greater on Sunday. Well, I wonder about this and so I'm going. On the whole they are rather different from the Jews I've known so far: 1.) because they keep a *simple* house 2.) because they are anti-semitic.[6]

In the gathering Bartók's intense shyness evaporated. He was swept along by the simple fun of the group.[7] There was no serious music-making, just good humour, tea, and social amusement, which lasted well into the evening. Bartók had rarely enjoyed himself so much, and was happy for an invitation to come again a fortnight later.[8] Unable to repay this hospitality in kind, Bartók expressed his gratitude through the tools of his trade. During the following week he sent two postcards to Adila, each containing a theme from *Don Quixote* by Richard Strauss, then one of his idols.[9] Two days later Adila received an original composition, an Andante, which was

[4] *BBcl* p. 60.
[5] Letter, Bartók to his mother, n.d. [12 Nov. 1902], *BBcl* pp. 74–5.
[6] Letter, Bartók to his mother, n.d. [19 Nov. 1902], *BBcl* p. 76.
[7] *BBcl* pp. 77–9.   [8] *BBcl* pp. 80–1.
[9] See unpublished postcards: Pierpont Morgan Library (25 Nov. 1902), MTA BA-B 4506 (27 Nov. 1902).

entitled 'In memory of 23 November 1902' (the day of the Arányis' party).[10] Further still, at about this time Bartók wrote a canonic Duo for violins, which he presented to Adila.[11]

As well as thanks, the gifts expressed Bartók's increasing affection for Adila. At its basis was an innocent sharing of musical pleasures, between good friends rather than lovers. As Joseph Macleod records in his biography of the sisters, Bartók was once teased to confusion during an Arányi family gathering about the nature of his affection for Adila. Playing the Hungarian game called 'how-like-you-this?', Bartók was flummoxed by the question 'how-like-you-Adila?'[12] The nature of their friendship can best be seen in Bartók's correspondence with Adila. On 29 December 1902, for instance, he wrote excitedly from his mother's home in Pozsony with the latest musical chit-chat:[13]

Thanks for your card. I am curious about your impressions of Ysaÿe! Just imagine what two Christmas presents I received were like: from Professor Gianicelli, a signed photograph of Richard Strauss, which was his property until now; and from Professor Thomán, the score of one of Strauss's earlier symphonic poems (Tod und Verklärung = Death and Transfiguration). This, also, was due to my success on Monday! I am unbelievably hard-working. The day before yesterday I finished composing my symphony and began to score it. I am learning a heap of piano pieces. I trust, my Lady, that you also are playing much music.

With greetings,

B[2]

Two days later he dispatched greetings for the New Year, with patriotic zeal inking out the German-language description of a Pozsony scene on the back of the postcard.[14] From the relationship Bartók received considerable musical feedback, learning from Adila of how Hubay (the leading violin teacher at the Academy) had praised him in a lesson,[15] and gaining the family's candid comments about his new work, an ambitious Study for the Left Hand.[16] Although Bartók enjoyed spending time with the Arányis he

---

[10] MTA BA-B 3341/a; facsimile, with appended notes by László Somfai (Budapest, 1980).

[11] MTA BA-B 3342; facsimile in Denis Dille, *Thematisches Verzeichnis der Jugendwerke Béla Bartóks, 1890–1904* (Budapest, 1974), 291.

[12] Macleod, *Sisters d'Arányi*, p. 23.

[13] Unpublished postcard, Bartók to Adila Arányi, 29 Dec. 1902, in Hungarian, SzFAC.I. Bartók probably refers to the Budapest concerts of the Belgian violinist Eugène Ysaÿe on 3 and 4 December. The 'success' concerns his own rendition of Strauss's *Heldenleben* before the Academy's professors (*BBcl* p. 323). The symphony mentioned is that in E flat major. Only the Scherzo movement of the work was orchestrated by Bartók, and subsequently performed. Bartók's closing 'B[2]' was used several times on postcards to Adila Arányi. Among other extant correspondences not discussed below are those from Bartók to Adila Arányi of 10 Jan. 1903, 26 Jan. 1903, and 25 Feb. 1903 (MTA BA-B 3805, 4555, 3553), and those of 10 Apr. 1903 and 31 Dec. 1903 (SzFAC.I).

[14] SzFAC.I (31 Dec. 1902).   [15] *BBcl* p. 81.   [16] *BBcl* pp. 84–5.

was careful not to fritter away the precious final months of his Academy course: '. . . tomorrow afternoon I go to the Arányis, but only for a short time for I have much work to do. Besides, it is now carnival-time, when it is risky to spend time in the company of the young since they easily start to dance.'[17] From a letter to his mother of 15 February 1903 it appears that Bartók also contemplated performing in public with Adila, but, for whatever reasons, decided against this collaboration.[18] On his way to a concert in Nagyszentmiklós (his birthplace), for which he had considered her as a partner, Bartók nevertheless sent Adila a postcard of the new synagogue in Szeged and told her of a long wait for a connecting train.[19]

The correspondence between Bartók and Adila continued for at least five years, during which time he was a frequent visitor in the Arányi home. On 30 November 1905, for instance, having just returned to Vienna from London, Bartók wrote to Adila giving her advice about arranging a concert in Pozsony, but also fuming at his own lack of success in the Paris Rubinstein competition earlier in the year:

Well, you also talk about my 'successes' in Paris. I would go into a rage at just hearing the name 'Rubinstein' should I not have acquired over the years some measure of imperturbability. It is really a story worthy of a comic journal that our country's papers have recorded the débâcle in Paris in the column headed 'Hungarian's Success Abroad'. Thanks for your congratulations, anyway. I am sure that you meant well. I take them as being for my success yesterday in Vienna.[20]

The two remained in close contact. When Adila's final examination for her Artist's Diploma approached later in 1906, Bartók took her to a music shop in Budapest and bought her all the violin concertos in stock.[21] As Adila recalled in a talk presented half a century later, the work which she received for limited-time preparation, the Dvořák Violin Concerto, Op. 53, had sadly not been among them, although Bartók made amends by helping her each day during the allowed week of preparation.[22] When the time arrived for an examination performance of Tartini's 'Devil's Trill', Bartók sent Adila a postcard covered with over 500 tiny faces. If they came to an even number, he proclaimed, then a good performance could be expected; if odd the reverse.[23] Whatever the number, Adila did gain the much-coveted diploma.

By late 1907 Adila was starting to shape an international career. She had played Beethoven's Violin Concerto, Op. 61 at her Viennese début, to wide

[17] Letter, Bartók to his mother, 17 Jan. 1903, *BBcl* p. 82.
[18] *BBcl* p. 88.   [19] SzFAC.I (11 Apr. 1903).
[20] Unpublished letter, Bartók to Adila Arányi, in Hungarian, 30 Nov. 1905, SzFAC.I. The 'success' in Vienna concerned the first performance of some movements from Bartók's *Suite* No. 1, Op. 3.   [21] Macleod, *Sisters d'Arányi*, p. 44.
[22] Unpublished paper, 'Béla Bartók', dated 23 June 1955, quoted in Macleod, *Sisters d'Arányi*, p. 44; unpublished letter, Adrienne Fachiri Camilloni to the author, 22 Oct. 1987.
[23] Macleod, *Sisters d'Arányi*, p. 44.

acclaim, and had since studied with her great-uncle Joachim in Berlin. Bartók, too, had advanced professionally, having joined the piano staff of the Budapest Academy of Music. Yet he was still prepared to turn down requests that they perform together, while continuing to indulge in informal music-making on Sundays.[24] In all these years it would appear that the two never collaborated in public.

Meanwhile Jelly's skills as a violinist were being honed under the guidance of teachers Vilmos Grünfeld and Jenő Hubay. In natural talent on the instrument she exceeded even her sister, and by 1908, at the age of 15, was performing jointly with Adila, and sometimes with Titi as well.[25] On 10 October of that year the three sisters combined to give a formal concert in Budapest featuring mainly Romantic violin works.[26] In early 1909 they ventured to England, where acquaintances of the late Joachim, including Donald Tovey, provided them with concert venues. (Tovey and Adila had both been at Joachim's side at his death in August 1907.) The first performances took place in a school hall at Haslemere in Surrey. Perhaps because Bartók had himself been to Haslemere in 1905 to visit the Oliversons,[27] the sisters sent him a short card telling of the splendid time they were having.[28] On 2 April Adila and Jelly took part in a Classical Concert Society performance in Newcastle-on-Tyne, in collaboration with Tovey.[29] The critics were impressed. A local correspondent of *Musical News* noted that Jelly's performance of Brahms's *Hungarian Dances* was 'remarkable for verve and emotional warmth, especially from such a youthful player'.[30] Socially, the sisters' acceptance was equally remarkable. The wit, charm, and social poise, which had so captivated Bartók, quickly earned them many English friends and helped them to secure extra engagements. In succeeding years they visited England repeatedly, taking part in the festival programmes at Haslemere, but also gaining contacts in many other parts of the country. By February 1913 the sisters' musical activities were self-supporting and so they decided to take rooms in London. Their mother lived with them, while their father remained at his police post in Budapest.

So successful and happy were these early days of residence in Britain, with their frequent concerts and many social outings, that the seriousness of the political situation in Europe escaped the sisters' attention. Only when unmistakable signs of war loomed did they realize the danger of becoming aliens in an enemy country. Too late, they tried to return to Hungary, only to find at Ostende that their continuing passage could not be guaranteed.

[24] See, e.g., SzFAC.I (18 Nov. 1907).
[25] Jelly Arányi was born in 1893. The date of 1895 which is found in many reference books may have been circulated by the family to promote recognition of Jelly as a child prodigy.
[26] Programme, BL Arányi Collection.   [27] BBcl p. 144.
[28] MTA BH 39.   [29] Programme, BL Arányi Collection.
[30] H. W. [H. Werner], 'In the Provinces', *Musical News* (London), 17 Apr. 1909.

Quietly they returned to Britain, entering via the British queue (where passports were not required) to avoid identification as aliens.[31] Now illegal immigrants as well as enemy aliens, the sisters kept a low profile and adopted the French version of their surname, d'Arányi. When eventually the police did manage to trace them, they were staying at the country home of Herbert Asquith, none other than the Prime Minister! Macleod, drawing on the recollections of Jelly, tells how Asquith drove the family to the police station for questioning.[32] Needless to say, they were not interned. Within a year the two elder sisters had married most respectably; Titi's husband was Ralph Hawtrey, a Treasury official, Adila's was Alexandre Fachiri, a lawyer of American origin living in London. Through the war Adila and Jelly continued to perform chamber music whenever the opportunity presented itself. Adila also took in pupils.

When the war ended there was no question of returning to Hungary. For reasons of marital links, the political situation, and economics, permanent residence in Britain was the only sensible course to follow. Gradually much of the concert life before the war was re-established and the sisters' talents, Jelly's in particular, were increasingly in demand. Not only could Jelly play superbly, but, with maturity, she had also developed a considerable stage charisma. After a Wigmore Hall recital in May 1920, a critic from *Time and Tide* commented:

Personality counts on the concert platform as much as elsewhere. To ignore its power is as stupid as to grow hysterical about it. Jelly d'Arányi has a great asset in her appearance. She looks like a picture by Alfred Stevens—at any rate, on this occasion she recalled his portraits, garbed as she was in a full-skirted pink silk dress, her dark hair confined by a pink ribbon—and her physical movement is as rhythmical as her playing.[33]

So, too, the *Lady* wrote of her playing during the following year: 'Vigour and animation are the most striking characteristics of Miss Jelly d'Arányi's violin playing. Quite evidently feeling and living the music she plays, this gifted artist carried her hearers away with her.'[34] Jelly was now widely recognized as one of those rare artists who combined technical brilliance with a compelling emotional warmth.

In November 1920, when Heseltine had first written to Bartók and suggested a visit to England, the Hungarian had immediately realized the possibility of some collaboration with the Arányis.[35] Bartók wrote to Adila, who had vaguely mentioned such an idea earlier in the year, to try to gain a more definite commitment,[36] but the Arányis took no action. It was only in

[31] Macleod, *Sisters d'Arányi*, p. 89.   [32] Ibid. 91–2.
[33] Christopher St John, 'Music of the Week', *Time and Tide*, 14 May 1920, p. 22. For an excellent photograph of Jelly see the Supplement to the *Strad*, no. 437 (Sept. 1926).
[34] Anon., 'Jelly d'Arányi', *Lady*, 1 Dec. 1921.   [35] See VBBH pp. 139–40.
[36] BBcl pp. 308–10.

the autumn of 1921, while all three sisters were holidaying in Budapest, that real progress towards a British visit was made. Having discussed concert plans with the elder two sisters, Bartók wrote to Calvocoressi in London recommending a closer collaboration with the family to further his interests there.[37] The letter concluded with a request that Calvocoressi find out, in confidence, if a performance with Jelly Arányi would be beneficial to his tour. Bartók knew that she gave many concerts in Britain, but he had not had the opportunity of hearing her play. He did not want to take part in a general chamber music programme with a player of inferior standard.

Within a month Bartók's concerns had vanished. One morning Jelly, who was not particularly enjoying the holiday, decided to visit the Bartóks. At this time Bartók was living in the Lukács family home on Gellért Hill.[38] As he had not heard Jelly play since her early teenage years, the two made music together. Bartók was stupefied. Musically their rapport was incredible; personally they were complementary, Bartók's dour nature balancing Jelly's effervescence. There was no longer any need for Calvocoressi's private investigations of Jelly's performing standard. 'I found her to be an excellent violinist, so that I would readily play with her publicly anywhere,' wrote Bartók soon afterwards to Calvocoressi.[39] Having earlier been unwilling to perform other composers' works during his British tour, he now assured Calvocoressi that he was more than willing to play such pieces with Jelly. In composition, too, Jelly had proven an immediate stimulus. Bartók hardly needed her suggestion of a violin sonata over a meal at the Gellért Hotel before setting to work.[40] Márta, Bartók's (first) wife, ever selflessly devoting herself to his interests, wrote to his mother with the news:

I have never yet received such a beautiful 'surprise' for my birthday! Ich bin ganz aus dem Häusel [I am quite beside myself]. Béla took me by surprise with the news that—he composes! Again at long last. I hardly contained myself this morning when, as a birthday present, he showed me the violin sonata on which he's working. No-one is to know about it until it's ready—but of course I had to write to you. He has allowed that. Well, could anyone get a more beautiful present than this? I am so happy. How afraid I was that all the deprivations of recent years would finally enfeeble Béla's ability to work—and how grateful I am to Jelly Arányi, whose wonderful playing on the violin has drawn out of Béla this (as he tells) long dormant plan. . . . So, I've now let out that part of my joy which was brimming over—now I am going to do the house-work, then I'll continue.[41]

---

[37] BBrCal pp. 205–7.
[38] For a short record of Bartók's home life at this time, see Mária Lukács-Popper, 'Bartók our House-guest', NHQ no. 84 (Winter 1981), 103–6.
[39] Letter, Bartók to Calvocoressi, 10 Oct. 1921, in German, BBrCal pp. 207–9.
[40] Macleod, Sisters d'Arányi, p. 136.
[41] Letter, Márta Ziegler to Bartók's mother, 19 Oct. 1921, BBcl p. 325.

Despite the birthday treat Bartók dedicated the sonata to Jelly, not Márta. By 9 November its first two movements had been completed, and Márta was diligently copying them out.[42] It was a sonata written on impulse, with Jelly's technical skills always in mind, and a piano-part equally tailored to Bartók's performing strengths. Although British plans were now being formalized, Bartók wished to work with Jelly more widely. In letters to her he begged that she should undertake a tour of Hungarian cities with him, or come to Frankfurt-am-Main or Paris when he would be there.[43] Jelly was initially receptive to these proposals, but only the Parisian excursion eventually took place.[44] By Christmas 1921 the technicalities of the sonata were being sorted out between the two, with Bartók seeking Jelly's advice on the playability of some of the more complex violin passages. Then staying with his sister in the country, Bartók was in a relaxed mood and told Jelly a little about the life there: the cattle, frogs, and water-beetles.[45]

When the concert schedule for Bartók's British tour of March 1922 was finalized, it involved strong participation from the Arányis. The opening concert at the residence of the Hungarian chargé d'affaires on 14 March featured both Jelly and Adila, its chief interest being the new violin sonata.[46] So, too, the public concert on 24 March and the closing concert at the Wilsons' home on 31 March centred upon Jelly's performances of Bach, Beethoven, Mozart, and Bartók sonatas. In a letter to her father, who was still working in Budapest, Titi gave a family view of the public concert.[47] Few people had really understood Bartók's music, she wrote, but nevertheless the concert had been a success, not least because of Jelly's fine playing. Although Bartók did not stay with the Arányis, he spent much time at 18 Elm Park Gardens, Chelsea, the home of Titi, where Jelly and her mother were then also living. After the most important London engagements were over, he stayed for a few days at an address in Half Moon Street, off Piccadilly.[48] Jelly's reminder to Bartók some days later to return the key of this flat to its owner suggests that the family helped in arranging this central London sojourn, too.[49]

More successful than he had expected on the musical front, Bartók was not making such headway in personal relations. The Arányis were beginning to find him tiresome. Matters came to a head in Paris, where Jelly had gone—chaperoned by her mother—to take part in several concerts with Bartók. They all stayed in the Hotel Majestic,[50] and the closeness and frequency of contact appears to have exacerbated existing tensions. Bartók

[42] Macleod, *Sisters d'Arányi*, p. 136.   [43] Ibid. 136–7.
[44] SzFac.IV (10 Dec. 1921).   [45] Macleod, *Sisters d'Arányi*, p. 137.
[46] Details of this and subsequent Bartók–Arányi concerts can be found in Chapters 3 and 4, above.
[47] SzFAC.IV (25 Apr. 1922).   [48] MTA BA-B 3477/233.
[49] MTA BA-B 3477/129.   [50] See MTA BH D-III.52 Számlák.

was altogether earnest in his pursuit of Jelly, but she could not take seriously his display of affection. Rather naïvely, she had not realized how her uninhibited enthusiasm in musical matters could be interpreted otherwise.[51] In Paris she soon found herself seeking to avoid Bartók, and she snubbed his invitation that they attend a concert together.[52] Their own performances were, however, still very successful. To her father in Budapest, Jelly wrote:

My darling Papa,

  How I long for you to be here, except that the weather is so bad, one can't enjoy anything—Last Saturday I had about the most exciting musical event in my life. Bartók and I played his sonata at a concert, and all the greatest living composers came to hear it; that was in the afternoon, and in the evening they all came to a soirée to hear it once more—Ravel, Stravinsky, Szymanowsky and many other less important ones. They all said I was the greatest violinist etc. I was very excited and pleased about it. . . . Bartók is going tomorrow—we are glad, as he is a little difficult to be with—, I must say he was a very great success both here and in London.[53]

Whatever the reasons, this 'little difficulty' turned quickly to a positive dislike. Writing a card to her sister Titi soon after Bartók had left Paris, Jelly confided:

My dear Csingi—I just can't say how glad we were for your letter today. B. had been making life here beastly for us. I'll write more, and at length, from L[ondon]. It was a waste of time to help B. so much—he's an awfully disgusting character. We're well otherwise. A million kisses. Tomorrow we go home. B. has already gone. Do write a lot—it's such a joy for us! Sái[54]

Bartók had gone too far, and what is more, taken his rejection badly. At a luncheon together he had drunk too much, causing further offence.[55]

  This emotional conflict between Bartók and Jelly Arányi was more than a 'storm in a teacup'. Bartók's performing opportunities in Britain were threatened. He had offended his chief supporters there, and could not expect them to go out of their way again to help him. Hopes of a visit to Britain in November quickly dwindled. When the concert agency Ibbs and Tillett tried to arrange a joint recital in Wales for Jelly and Bartók, Jelly replied that he was not coming to Britain, and suggested that the agency contact Bartók himself if it wished to pursue matters further.[56] In a letter to Calvocoressi of late June Bartók bemoaned the lack of any news from the sisters, and its consequence: 'no matter how I wish to return to England next winter, I cannot realize this plan for a while'.[57] By August there had

[51] See Macleod, *Sisters d'Arányi*, pp. 139–40.   [52] SzFAC.IV (6 Apr. 1922).
[53] Unpublished letter, Jelly Arányi to her father, 13 Apr. 1922, in English, SzFAC.IV.
[54] Unpublished card, Jelly Arányi to Titi (Hortense) Hawtrey, n.d. [c. 18 Apr. 1922], in Hungarian, SzFAC.IV. 'Csingi' and 'Sái' were two of many nicknames used between family members.   [55] See Macleod, *Sisters d'Arányi*, p. 139.
[56] SzFAC.IV (4 May 1922).
[57] Letter, Bartók to Calvocoressi, 24 June 1922, in German, BBrCal pp. 212–15.

still been no word from the Arányis, causing Bartók to write to Calvocoressi with a further lament:

Now I try to practise the sonata here with Waldbauer and the young violinist Székely; I hope that both of them will play it quite respectably; but—as I well know the playing of both—I am, alas, quite sure that neither of them will be able to perform the piece with anything like that wonderful perfection which we were lucky to hear from that unique artiste.[58]

Time passed. By March 1923 plans for a British visit had still not advanced. Bartók would have liked to have performed his new (Second) Violin Sonata (1922) with Jelly Arányi, to whom he had dedicated it in spite of his rejection. But with no prospect of such a collaboration, the first performance of the work had taken place in Berlin on 7 February, with Imre Waldbauer joining the composer.[59]

Soon afterwards, through the activity of the International Society for Contemporary Music, a tour of Britain became possible. At the request of this society Jelly agreed to perform with Bartók at a concert on 7 May featuring both the violin sonatas. Her contribution to the performance's success was acknowledged by the critics, although the acerbity of Bartók's touch detracted somewhat from the ensemble. Writing to her father in Budapest, Titi again provided the family view of Bartók's stay in London: 'Bartók goes home today. He scored a success and made some money as well. He behaved completely correctly and was more congenial. We, along with others, consider he is a great master, though he is still little understood.'[60] Bartók was rehabilitated, albeit with some caution, to the position of family friend and 'elder brother'. Thirty years later Adila would recall his innocent affection for fruit during this visit. Before dinner he would often help himself to several pieces from the fruit bowl, and even took a large supply of bananas back to Hungary as they were hard to buy there.[61] During Bartók's first visit of 1923 the sisters were most concerned about the rapidly deteriorating health of their mother. Jelly could not think of Bartók, or anyone else, at this time of sadness. Within a month of his departure her mother died. Despite their personal grief the sisters continued to perform at an increasing frequency.

When Bartók revisited Britain late in 1923 his concert connections with the sisters were further repaired. Jelly performed with him twice, on 30 November at a public Aeolian Hall concert featuring the Second Violin Sonata, and also at a private function. Adila, who had previously taken a secondary role during Bartók's visits, now appeared with him, probably on

---

[58] Letter, Bartók to Calvocoressi, 20 Aug. 1922, in German, BBrCal pp. 215–16.
[59] BBrCal pp. 217–18.
[60] Unpublished letter, Titi (Hortense) Hawtrey to her father, 14 May 1923, in Hungarian, SzFAC.IV.   [61] Macleod, *Sisters d'Arányi*, p. 139.

three occasions, including a chamber music concert held at her own home in Chelsea. Starting at 9 o'clock in the evening this concert featured a mixed programme of Bartók piano solos and folksongs (sung by Dorothy Moulton), and Beethoven's 'Archduke' Piano Trio, in which hosts Alexandre Fachiri (cello) and Adila (violin) joined their guest to conclude the entertainment. For the select audience a printed programme was issued, with notes by Calvocoressi, and a cover charge of 10s. 6d. was levied to defray costs and Bartók's fee.[62] By this time the Fachiris' musical evenings were renowned for their mixture of good music and social gaiety. As Macleod has recorded in his biography of the sisters:

The gatherings were anything but solemn. If Beethoven were to be played, he was played respectfully and well; but afterwards he was in shirt sleeves, especially if anyone had been giving a public concert and needed to let off steam. So you might see Harold Samuel dressed up as Queen Victoria singing comic songs at the piano.[63]

Harold Samuel, perhaps, but hardly Bartók!

The extent to which relations had healed is shown in the willingness of Jelly and Adila to join Bartók in Geneva for a concert on 20 December 1923. The main items for this performance were the sisters' *pièce de résistance*, Bach's Double Violin Concerto in D minor, and Bartók's First Violin Sonata. Jelly and Bartók performed this second work, although Adila had also played it with Bartók in London during the previous week. Jelly foresaw some trouble with the 'very snobbish audience'.[64] While the Bach performance was judged a huge success, the Bartók sonata drew hardly any applause. It had been 'too difficult a morsel for the audience', Bartók concluded with resignation.[65] Having seen more favourable receptions, the sisters felt genuinely sorry that his music was so hated by the conservative Swiss.[66] On departing they assured Bartók that in future years they could gain him many engagements in England on a regular basis.[67]

Bartók was not to test the validity of their assurance. Rapidly he moved out of the British orbit, losing many contacts so painstakingly established.[68] Teaching-work, the preparation of his large study of Romanian folk-music, new family responsibilities, and illness all combined to make international concert trips and creative composition all but impossible for several years.[69] Whether Bartók wished it or not, any continuing association with Jelly would clearly have caused anxiety for his new wife, Ditta. The Arányis, for their part, also did not push for further concerts involving Bartók. Jelly was at the height of her powers, inspiring Ravel's *Tzigane* in 1924 and Vaughan

[62] MTA BH 2049/130.
[63] Macleod, *Sisters d'Arányi*, p. 243.   [64] SzFAC.IV (19 Dec. 1923).
[65] *BBcl* p. 347.   [66] See SzFAC.IV (c. 21 Dec. 1923).
[67] *BBcl* p. 347.   [68] BBrCal pp. 221–4.
[69] See Béla Bartók, jun., 'Béla Bartók's Diseases', *Studia Musicologica*, 23 (1981), 434; Malcolm Gillies and Adrienne Gombocz, 'The "Colinda" Fiasco', *ML* (forthcoming).

Williams's *Concerto Accademico* in the following year. Bartók was only one of many to be fascinated by her talents. Although very occasionally Jelly did perform one of 'her' Bartók sonatas—on 9 June 1925, for instance, before the Oxford University Musical Club[70]—the sisters did little else to promote Bartók's music in Britain. Some musical contact with the family might have been re-established in October 1927, when Bartók next came to Britain, were it not that Dezső Rácz, who paved the way for Bartók's first BBC engagements, had himself suffered an embarrassment at the hands of the sisters when a student in Budapest.[71]

It was not until December 1927 that Bartók and Jelly Arányi met again. Both were on extensive concert tours of the United States, Jelly having, in addition, several important recording contracts to fulfil. Soon after Bartók had arrived in New York they were, independently, invited to the same party. (This was probably on 18 December at the Bohemians Club, where Mengelberg, Rakhmaninov, and Szigeti were also present.[72]) Writing just after Christmas to Titi, Jelly noted: 'I saw Bartók again at a party—he looks so unhappy, but his music, played by him again, is a great thing.'[73] Bartók's available correspondence makes no mention of the meeting, nor of an important change in his concert arrangements. Szigeti and Bartók had been booked to contribute several items to an afternoon concert on New Year's Day 1928, organized by the Society for Contemporary Music in Philadelphia. Late in December Szigeti received a prestigious offer to perform in Chicago on this same day. Who else could stand in for him at such short notice in a performance of Bartók's Second Violin Sonata? Immediately Szigeti was on the telephone to the work's dedicatee. Jelly later recalled to Titi:

. . . he [Szigeti] begged me to take his place—there was no fee, only expenses. I was then still very low and nervy and told him I could not possibly do it, as I have only Saturday to practise the work, and I will be too tired after recording 3 whole days etc.; however he begged me so much, and I like him, so I gave in. I felt better at once, because I was doing [*sic*] a real sacrifice, and I was a little bit proud of myself.[74]

Bartók certainly did not mind the change. It was an unexpected chance, after four years, to play this work again with the one for whom it had been fashioned. On the train to Buffalo on the day after the concert, Jelly wrote of the personal side of their reunion:

. . . Philadelphia was quite unimportant, like the Contemporary Music Society in London. But it was nice to be with Bartók! You can't imagine what it means in this country to meet old friends—even B.! Especially as all the unpleasantness is gone. I

[70] With Ernest Walker as associate artist. See BL Arányi Collection.
[71] See *DB*.iii pp. 132–4.
[72] Béla Bartók, jun., *Apám életének krónikája* (Budapest, 1981), 255.
[73] Unpublished letter, Jelly Arányi to Titi (Hortense) Hawtrey, n.d., in English, SzFAC.IV.
[74] Ibid.

can only remember I knew him as a child. He was so happy to be with me, as he is very homesick—I personally am happy now— . . .[75]

Despite their personal affability in Philadelphia, the duo met considerable opposition from the floor of the auditorium. The concert was judged too long and dissonant by the contemporary music enthusiasts of the city. Although Szigeti's cancellation 'because of an infected finger' had already been notified, the hall was full to overflowing with people curious to fit sounds to the much-discussed names of Ornstein, Ravel, and Bartók.[76] While works by the first two composers gained some degree of sympathetic response, the two works by Bartók met stiffer opposition. Commented the *Philadelphia Inquirer*:

The rush to get in, however, was only exceeded by the rush to get out after Yelly [*sic*] d'Arányi, Hungarian violiniste, and the third featured figure of the afternoon, had wrung and wrested cries of anguish from her forlorn fiddle in Bartók's Sonata for Piano and Violin, No. 2, with the composer aiding and abetting the tonal torture at the piano. The performance left some members of the audience apparently stunned, but one piano teacher, hurriedly departing, was heard to say, 'Why, my ears positively hurt!'

Jelly's superb artistry in the work, and also in Ravel's *Tzigane*, was none the less acknowledged by the reviewers, although Bartók's piano-playing attracted little attention. As so often in the past, his music was recognized as the most original, and also the ugliest, in the programme.

This was the last formal concert in which Bartók performed with a member of the Arányi family, although he did remain in contact with its members. When on a trip to Budapest in the late summer of 1928, Adila visited Bartók at the Academy of Music and found him 'so pleased and nice'.[77] In early 1930 Bartók suggested Jelly as associate artist for a BBC studio performance later in the year.[78] His more recent associate, Zoltán Székely, was named as a second choice. Three months later, when nothing definite had been arranged by the BBC, he again pushed for a prompt answer, suggesting also the possible involvement of the Hungarian singer Mária Basilides, who would be in Britain at that time.[79] Eventually a late evening broadcast was arranged for Bartók and Jelly Arányi over the National station on 24 November 1930.[80] The main item, Mozart's Violin Sonata in A major, K 305, was flanked by short Bartók and Kodály piano

---

[75] Unpublished letter, Jelly Arányi to Titi (Hortense) Hawtrey, 2 Jan. 1928, in English, SzFAC.IV.

[76] See reviews: S. L. L., 'Ornstein Quintet Given Premiere', *Public Ledger* (Philadelphia), 2 Jan. 1928; Linton Martin, 'Modern Music has a New Year Party', *Philadelphia Inquirer*, 2 Jan. 1928. See also *Philadelphia Record*, 2 Jan. 1928.

[77] Unpublished letter, Adila Fachiri to Titi (Hortense) Hawtrey, n.d. [about Sept. 1928], in English, SzFAC.IV.   [78] *BBlev*.v p. 370.

[79] Ibid. 376.   [80] MTA BA-B 2400/46.

pieces. At home, Bartók appears to have kept quiet about this performance with Jelly. The postcard written to his wife on the day mentioned much rehearsal, but failed to say with whom;[81] the concert list kept by Bartók's mother makes no mention of another artist being involved on this day;[82] and the recent chronicles of Bartók's elder son, drawing on family sources, even record Mária Basilides as co-artist.[83] On the following day Bartók dined with the Fachiris, writing afterward to his wife: 'Yesterday evening I had dinner with the Fachiris (= Adila); there were no other guests. I ate all sorts of exotic fruit which they had ordered just for me (this was the most noteworthy event so far).'[84] Links with the Fachiris were further strengthened in March 1932, when Bartók stayed at their London home for several days.[85] But here, after thirty years, the documentary trail fades. Bartók probably saw some members of the family again, but there is no conclusive evidence available from sources on either side of the relationship. Their paths diverged. The sisters were to gain much publicity later in the decade over the performance of the Schumann Violin Concerto in D minor (1853), and the supposed assistance of supernatural forces in the work's rediscovery.[86] Death, too, intruded into the family's circle. Alexandre, Adila's husband, died in 1939, and their lifelong friend Sir Donald Tovey followed in 1940, preceding Bartók by five years.

During the 1950s, while going through memorabilia, Adila came across a few sheets of card in the form of a manuscript. They dated from the turn of the century. Here was an Andante entitled 'In memory of 23 November 1902', a recollection of a happy family party shared by a shy 21-year-old student, Béla Bartók. Although then nearly 70 years of age, Adila none the less gave the much-delayed première of the work at one of her last Wigmore Hall recitals on 6 July 1955. It was a 'spotless *morceau de salon*', claimed Donald Mitchell in the *Musical Times*, 'the perfect stuff for a guessing game. Strauss might come to mind, and even Mendelssohn, but Bartók never.'[87] This was a fitting, final tribute from the Arányis to a lost friend, by then recognized as one of the century's greatest composers.

[81] BBcl p. 496.

[82] Mrs Béla Bartók, sen., 'Béla hangversenyeinek jegyzéke', MS in the possession of Béla Bartók, jun. (Budapest), p. 14.

[83] Bartók, jun., *Apám életének krónikája*, p. 296, and *Bartók Béla műhelyében* (Budapest, 1982), 195.   [84] Letter, Bartók to his wife, 26 Nov. 1930, BBcl pp. 497–8.

[85] BBcl pp. 525–6, 526; SzFAC.I (3 Feb. 1932).

[86] See Baron Erik Palmstierna, *Horizons of Immorality* (London, 1937).

[87] 'London Music', MT 96 (1955), 485.

EPILOGUE

By 1938 Bartók's physical presence in Britain was no longer needed to trigger a display of interest in his music. Over the preceding two decades he had become a significant enough figure in British musical life and had gained sufficient supporters for a regular, if not excessively frequent, schedule of performances, recordings, commentaries, and advertisements to be assured. In his absence from the country after 1938 the BBC and the International Society for Contemporary Music's local section continued to programme his works, even in the darkest days of the war. Promotion of his music was undertaken more systematically than ever by his new publisher, Ralph Hawkes, who, of course, had a vested interest in the matter.[1] A new generation of musicians was supporting Bartók's cause by including his works more frequently in London's concert programmes. The young Hungarian pianist György Sándor, who had studied with Bartók, performed the *Petite Suite* (1936) on 23 June 1938 during a Wigmore Hall recital. Some months later his compatriot Andor Földes was more daring and performed Bartók's piano Sonata (1926) at his London début.[2] Critical wrath was largely forestalled by its careful placement among works by three other great Bs of musical composition: Bach, Beethoven, and Brahms. During the war years, further promotion of Bartók's music would be undertaken by such musicians as Louis Kentner, Ilona Kabos, and Yehudi Menuhin.

In the orchestral arena, Sir Thomas Beecham performed the *Music for Strings, Percussion, and Celesta* (1936) before a Queen's Hall audience at a Royal Philharmonic Society concert on 30 March 1939. Although the work had been presented by the BBC in January of the previous year, this mass public performance in a mainstream concert series displayed its virtues more widely. For many critics, including the ever-sceptical Richard Capell, the performance raised the prospect that perhaps Bartók could write truly great music.[3] The harsh austerity of so much of Bartók's earlier output had been sacrificed a little, and to good purpose, Capell surmised. British gramophone enthusiasts of 1939 were also able to hear a wider range of Bartók's music, with the release of recordings of the *Three Rondos on Folk Tunes* (1916/27) and *Romanian Folk Dances* (1915) featuring the pianist Lili Kraus, another of Bartók's pupils.[4] The records were well reviewed from musical and

---

[1] For letters of this period from Ralph Hawkes to Bartók see MTA BH 130–75 and 186–95. Other correspondence, in both directions, is held by Péter Bartók (Florida) in collections 4711–4.

[2] See Andor Földes, 'Rehearsing with Bartók', *NHQ* no. 81 (Spring 1981), 24–5.

[3] 'Music of the Four Worlds', *DT* 31 Mar. 1939.   [4] Parlo-Odeon, R20434–5.

technical perspectives. The *Gramophone*'s critic concluded about this *12s.* release: 'As a tonic against crooners and swingers *et hoc genus omne*, this open-air music, redolent of the soil, is worth a guinea a record!'[5]

Over the years Britain had become for Bartók a haven of security, in a world of disintegrating social order. Having during 1938 entrusted many of his most precious manuscripts to friends in Switzerland, he came to realize that they might still be vulnerable there. With Hitler's invasion of Czechoslovakia in March 1939, Bartók felt compelled to relocate this collection. He was not convinced that the German dictator would respect Switzerland's neutrality. To Annie Müller-Widmann in Basle he therefore wrote on 8 April:

These last events (you know by now what I mean) appear to me alarming and dangerous. (I am convinced that you are all very concerned also.) With a heavy heart I feel that I ought to take other precautionary measures. And so I ask you to send my manuscripts along with their list to my publisher, Boosey and Hawkes, 295, Regent Street, London. Don't send them all together, but divide them among four or five or even more packages, sending one after the other, on different days. . . . London, after all, is further away from the land of horrors.[6]

But not far enough. Bartók recognized that in the coming war little could be taken for granted. Even in writing this letter he had acknowledged: 'The manuscripts should really go to America, but at present I don't have anyone there.' When the matter was raised with Hawkes he agreed for the manuscripts to be held at the New York office of his firm.[7] After a short British residence, Bartók's treasures were therefore forwarded across the Atlantic.

With the declaration of war in September 1939 little initially changed in Bartók's relations with Britain or the reception of his music there. The jingoistic fervour of 1914 was absent; music by foreign composers was not now subjected to bans in the concert-halls or publishing-houses. Bartók's position in relation to Britain was curious, as he was a disaffected citizen of a hostile country which had, none the less, declared itself 'non-belligerent'.[8] The mail continued to flow between Britain and Hungary, carrying much material concerning the publication of new scores, in particular the *Mikrokosmos* volumes. In December Bartók received the usual Christmas cards from old British friends such as the Wilsons.[9] The International Society for Contemporary Music even attempted to continue activities in Europe during this period of 'phoney war'. Bartók, as a member for 1939–

[5] A. R., 'Analytical Notes and First Reviews', *Gramophone* 16 (1938–9), 427.
[6] Letter, Bartók to Annie Müller-Widmann, 8 Apr. 1939, original in German, reproduced in Hungarian in *BBlev*.v p. 621.
[7] See Victor Bator, *The Béla Bartók Archives* (New York, 1963), 13–14.
[8] For a summary of political and economic changes in Hungary at this time see I. T. Berend and G. Ránki, *Hungary: A Century of Economic Development* (Newton Abbot, 1974), 167–82.   [9] MTA BH 2005.

40 of the Society's four-man Council, headed by Edwin Evans, was involved in plans for a Council meeting in Budapest in December 1939 and an annual ISCM Festival in the city during the summer of 1940. The former meeting had to be cancelled, but only in a letter to Bartók of 1 February 1940 from the Society's Secretary, Edward Clark, was the impossibility of holding the Festival finally acknowledged.[10]

As the war intensified through 1940, Bartók's British contacts became restricted to his publisher alone. Transport and communications became less reliable; Bartók himself resolved to leave Hungary, and by October had settled in New York. But in this same month, while important battles were still being fought over the skies of Britain, the issue of performance of Bartók's music was being canvassed among BBC music staff. Having surveyed a score of Bartók's *Divertimento* (1939) for string orchestra, an official in the Deputy Director of Music's office was moved to marvel that 'unlike most of the recent Bartók, it is not extraordinarily difficult either to hear or to play, and yet is characteristic of the composer . . .'.[11] The memorandum continued: 'It might, in fact, be the long hoped for work which will make the mature Bartók available, not only to the general musical public but also to the general orchestra.' The BBC decided that a performance was justified.[12] So too, early in 1942, did some recently published Bartók choruses find a ready advocate in the BBC.[13]

As new Bartók scores were now emanating from Boosey and Hawkes, it is not surprising that the first performance of his Concerto for Two Pianos, Percussion, and Orchestra (the orchestral version of his similarly titled sonata) took place in London. This performance, on 14 November 1942, preceded the Bartóks' own performance of the work in New York by two months. The expatriate Hungarian piano duo, Louis Kentner and Ilona Kabos, was joined by the London Philharmonic Orchestra, conducted by Sir Adrian Boult. Critical reception of the work was unenthusiastic. Comparing the two versions of this composition, *The Times*'s critic observed:

[The] balance of forces has been upset by the intrusion of the orchestra. The percussion now sounds like an excrescence on the orchestral texture and the pianos (at any rate in the Albert Hall) no longer sustain a clear argument. In giving the work a wider currency, which was his purpose in scoring it, the composer has overrun the limitations which were the source of its strength.[14]

[10] MTA BH 1487.

[11] BBC memorandum, 29 Oct. 1940, BBCWAC 47796.

[12] See BBCWAC 47796 (12 Apr. 1943). Community performances of the *Divertimento* soon took place: in Liverpool on 2 October 1940 with Louis Cohen's Merseyside Chamber Orchestra; in London on 22 May 1941 with the Jacques Orchestra.

[13] BBCWAC 47796 (3 Apr. 1942).

[14] Anon., 'Royal Philharmonic Society', *The Times*, 16 Nov. 1942. Kabos had also taken part, along with Eda Kersey (violin, and Frederick Thurston (clarinet), in the first British performance of *Contrasts* on 23 September 1942 at the Wigmore Hall.

Realizing an element of truth in this view, Kentner and Kabos repeated the work, now in its original unorchestrated form, during a Wigmore Hall concert in the following March. The composer Lennox Berkeley, who was then working for the BBC, felt moved to write to the chief planner of music programmes, Julian Herbage, urging that the sonata be included in a broadcast.[15] He hailed the performance as 'truly magnificent', and the composition as 'one of the most important pieces of recent years'. When the work was tentatively scheduled for a broadcast at the prime time of 7.15 p.m. on a Saturday evening, Herbage moved quickly to prevent this happening.[16] He recalled the shock to the audience at the work's performance by the Bartóks in 1938. In his opinion it needed to be given a less prominent placement in the broadcasting schedule, or, preferably, a performance of a more acceptable Bartók work—the *Divertimento* or *Music for Strings*—should be repeated.

It was not one of these compositions, however, but rather the (Second) Violin Concerto (1938) which proved to be the 'long hoped for work' of mass appeal. As an anonymous reviewer commented in *Cavalcade* in September 1944, Bartók was less known to British music-lovers than either Kodály or Dohnányi, because no major work of his had yet taken the public's fancy.[17] With the first British performance of the Violin Concerto on 20 September 1944, over five years after its world première in Amsterdam, the amendment of this state of affairs was finally underway. Boosey and Hawkes took care to prepare the public through several press articles, including a long essay by the American composer Henry Cowell in the house journal *Tempo*.[18] Cowell accurately predicted that the good blend of feeling and intelligence evident in the work would ensure 'an increasing and long-lived success'. The soloist was Yehudi Menuhin, who was then on a tour playing to troops and factory-workers. He introduced the concerto in a concert broadcast from Bedford, the wartime base of the BBC Symphony Orchestra.[19] By now a personal friend of the composer, Menuhin felt moved to send Bartók a telegram after the concert:

FIRST PERFORMANCE OF CONCERTO IN BRITAIN WITH SIR ADRIAN BOULT AND BBC ORCHESTRA EXCELLENT AND PROVED TREMENDOUS SUCCESS STOP RADIO PUBLIC AS WELL AS INTERNATIONAL AUDIENCE COMPLETELY CARRIED AWAY BY WORK STOP YOU ARE DEEPLY ADMIRED AND LOVED HERE STOP AFFECTIONATE REGARDS = YEHUDI MENUHIN[20]

---

[15] BBC memorandum, Berkeley to Herbage, 8 Apr. 1943, BBCWAC 47796.
[16] BBCWAC 47796 (2 Apr. 1943).
[17] Anon., 'Music: Bartók's Violin Concerto', in reviews collection BBCWAC 47796.
[18] 'Bartók and his Violin Concerto', *Tempo*, no. 8 (Sept. 1944), 4–6.
[19] See Adrian C. Boult, *My Own Trumpet* (London, 1973), 120–1.
[20] Unpublished telegram, 22 Sept. 1944, in English, MTA BH 195/5. Hawkes sent a similarly ecstatic telegram to Bartók on 21 Sept. 1944 (Péter Bartók collection 4714/118).

Not all reviewers agreed with Menuhin, of course. *The Times*'s critic, while admitting an original and justified form to the concerto, found it distressing: 'It must be the most violent violin concerto in existence and is like nothing so much as a man stopping a tank with a rapier.'[21] In the *Daily Telegraph* it was considered too 'fantastic and ostentatious' to be great music. But these critics were out of step with the public reaction. Letters poured into the BBC, prompting Ralph Hill to write in the *Radio Times* that the performance had proven 'without hesitation . . . one of the most important musical events since the beginning of the war'.[22] He went on to hail Bartók as 'a composer whose sincerity and skill are universally agreed to be beyond praise'. The other BBC organ, the *Listener*, gave the more personal impression of William McNaught:

Like, I suspect, a good many other people I have usually had to crane my neck to get in touch with Bartók. Admiration has come, but not without a little self-coercion. This time I had the novel sensation of thoroughly enjoying his music without having to try. . . . If the composing of concertos and the playing of rugger were much the same sort of thing, I should say that Bartók's game was full of good open play among the halves and threes. In too much modern music the ball is muddy and seldom leaves the scrum. This activity, which is of thought and device, is nothing new in Bartók; but here it has an unaccustomed geniality and sparkle. . . . On the whole, he shows himself a better mixer than we took him for. The concerto gives support to those who have maintained through sundry hard times at Queen's Hall and elsewhere that Bartók was, in spite of all, a good composer.[23]

At the BBC the need to capitalize on this wave of interest was realized. The concerto was again broadcast with the same personnel over the Home Service on 8 October. A recording of *Contrasts* (1938), made in 1940 by the clarinettist Benny Goodman, together with József Szigeti and the composer, was broadcast on 15 October, followed by a recording of the First String Quartet, Op. 7 one week later. But at the highest management levels a fear soon took hold that Bartók was now being over-represented. When on 12 October Sir Adrian Boult wrote to his successor as Director of Music, Victor Hely-Hutchinson, suggesting that more works by Bartók be programmed,[24] the response was cautious. To Julian Herbage, Hely-Hutchinson scribbled the note: 'May we speak on this? I am a little nervous of "feeding people up" with Bartók by excessively continuous propaganda, but we obviously want to watch suitable opportunities of representing him.'[25] Despite his

---

[21] See reviews: Anon., 'Bartók's Violin Concerto: Mr. Menuhin's Tour', *The Times*, 22 Sept. 1944; F. B., 'The World of Music: A Virtuoso's Concerto', *DT* 30 Sept. 1944.
[22] *RT* 6 Oct. 1944.
[23] 'Bartók', *Listener*, 28 Sept. 1944. See also his column 'Round about Radio', *MT* 85 (1944), 311.   [24] BBCWAC 47796 (12 Oct. 1944).
[25] Handwritten note on memorandum, Boult to Director of Music, 12 Oct. 1944, BBCWAC 47796.

earlier reticence about the Sonata for Two Pianos and Percussion, Herbage strongly supported Boult. He even went further by suggesting not only a revival of the *Dance Suite*, but also performances of the Second Piano Concerto (1931) and the recently revised *Suite* No. 2, Op. 4.[26] No such performances appear to have been arranged immediately, however. In the press, the BBC's 'feeding people up' with Bartók's music was already being noted. W. R. Anderson, in the *Musical Times*, queried the organization's current purpose:

If you are going to argue with those who dislike Bartók, it is useless, as the B.B.C. so scrappily does, to engage in a general argle-bargle now and then: even if its arguments were good, which they often are not, few of the opponents would believe them. . . . I believe that in trying to persuade dissidents to make distinctions we must ourselves begin by making much finer and keener and more efficient distinctions between these publics. Some probably can be saved; others, I think, cannot. This is what few musicians will allow. They are by nature wholesale missionaries, and they achieve perhaps as much as do the similarly wholesale would-be disseminators of religion (including martyrdom, now and then).[27]

The ascent of Bartók's fortunes was not significantly affected by these debates. In the early months of 1945 the pianists Noel Mewton-Wood and Ilona Kabos gave frequent performances in London of the Sonata for Two Pianos and Percussion. Ralph Hawkes was arranging a series of performances of Bartók's works for late 1945, which included four renditions of the Violin Concerto with Max Rostal and the Liverpool Philharmonic Orchestra, under Malcolm Sargent; the British première of the *Concerto for Orchestra* (1943), under Sir Adrian Boult's direction; and a 'little Festival' at the Wigmore Hall, devoted to the six Bartók string quartets.[28] Menuhin had plans to perform and record the Violin Concerto in London during November 1945, and the Hallé Orchestra had included the work in its winter programme.[29] America was also witnessing the first signs of a real wave of popularity: in August 1945, for instance, Menuhin played the concerto before an audience of 18,000 at the Hollywood Bowl, and he was already playing the Sonata for Solo Violin (1944), which he had commissioned from Bartók, before gatherings of troops. On 16 August 1945, having recently returned from London, Menuhin wrote to Bartók with the assurance: 'You have many admiring friends in London and a general public more than ready and eager to welcome you and your music. Boosey and Hawkes will certainly be right with you in promoting your works . . .'[30] With enthusiasm he expressed a desire only realizable in the

---

[26] BBCWAC 47796 (23 Oct. 1944).
[27] 'Round about Radio', *MT* 85 (1944), 373.   [28] *DB*.iii pp. 261–2.
[29] See, respectively: *DB*.iii p. 263; Michael Kennedy, *The Hallé Tradition* (Manchester, 1960), 329.
[30] Letter, Yehudi Menuhin to Bartók, 16 Aug. 1945, in English, *DB*.iii p. 263.

new era of peace: 'Our most ambitious hope is to do it [the Violin Concerto] with you in the audience, in *Budapest*!'

Menuhin's hope would never be realized. Shortly before noon on 26 September 1945 Béla Bartók died in a New York hospital, a victim of leukaemia. Death had cheated him by a few weeks of witnessing a British acclaim more widespread than anything seen in his lifetime. Because of the increasing press and radio attention to his music over the last year, his passing was widely noted and sparked a spate of obituaries and memorial essays. These writings neatly summarized the diverse British assessments of his activities, not just in composition, but also in the fields of performance, ethnomusicology, and pedagogy. Although many of the traditional stances of criticism were maintained in these memorial essays, forthrightness of opinion was noticeably less evident than in the past. This was not just out of respect for the dead, but because of the genuine quandary that death had caught Bartók at a time when a serious re-evaluation of his talent had been in progress.

First off the mark was *The Times*, with a short, unemotional obituary on the morning after Bartók's death.[31] There, the continuing influence of Gray's *Sackbut* article of 1920 is seen in the statement that Bartók's works were not bound by any accepted rules, certainly not rules of form. On the following day a longer, updated obituary appeared in the newspaper.[32] The recent Violin Concerto was now considered, but with an even-handed, noncommittal treatment which verged on meaninglessness:

The violin concerto, like the earlier piano concertos, raised antagonism in some minds by its asperity, but confirmed the impression made by all his music, even the most astringent, unaccommodating, and apparently perverse, that his compositions are perfectly logical, and therefore intelligible, if his premises, which are admittedly unusual, strongly personal to himself, and quite deliberately adopted, are accepted.

In the *Sunday Times*, Ernest Newman, by now in his late-seventies and the only critic who had heard Bartók consistently since his earliest, Manchester performances, identified a new compositional departure in this concerto with the writing of music which could readily be approached by the ordinary listener.[33] Yet to his mind there was no guarantee of a vital after-life for Bartók's works. 'To what extent', the aged Newman ruminated, 'did the conscious theorist in him pin down the wings of the unconscious creator?' A month later, after hearing the memorial performances of the six string quartets promoted by Boosey and Hawkes, Newman was a little more positive about the chances of survival for Bartók's works, but still harboured doubts as to the mass aural acceptance of them.[34] The most basic

---

[31] Anon., 'Death of M. Béla Bartók', *The Times*, 27 Sept. 1945.
[32] Anon., 'M. Béla Bartók: Hungarian Composer', *The Times*, 28 Sept. 1945.
[33] 'Musical Chart-Reading', *ST* 7 Oct. 1945.
[34] 'The Bartók Quartets', *ST* 11 Nov. 1945.

problem was that the beautiful geometric shapes in Bartók's music sometimes translated into 'repulsively ugly' actual sounds. This could be because Bartók's ear was years in advance of his time, or rather, as Newman's column ended, because his pursuit of tonal abstractions had led him to forget that for the audience music has a physical side as well as a geometric one.

The *Musical Times* surveyed Bartók's life-work at some length, and concluded that his influence on the contemporary musical world had to be classed with that of Schoenberg and Stravinsky, despite the more secretive and less obtrusive nature of his art.[35] It was left to Ernest Chapman, in *Tempo*, to speak of the qualities of the man as well as the musician.[36] He recalled the sight of this frail, selfless, shy man, 'of absolute moral integrity', 'a good "European" to his finger-tips', who preferred exile to the compromise of principles. To Chapman the unemotional, business-like face which Bartók presented to the world was merely a mask worn by a sensitive man as a defence against a hostile world. Here was a man ahead of his time, whose qualities would be seen more clearly as the world caught up with him.

The weeks passed, the many planned Bartók performances took place, and the debate on Bartók's significance intensified. In the dawn of the new atomic age an increasing number of younger listeners were prepared to accept the harsh sounds of his works. A battle between progressive and conservative forces was raging in British society at large, and music was by no means excluded from its ambit. Even death provided no truce in such a struggle. Writing to *The Times* on 15 November 1945, Louis Kentner observed:

How significant that even the death of Bartók has ... merely served to stimulate partisanship for and against his music, as if it had been yet another 'discordant' opus of this extraordinary man. Bartók to the last remained a rebel, and a seeker perhaps rather than a fulfiller. At 64 his life's work was not harmoniously rounded off but tragically ended with the land of promise just in sight. . . . Perhaps it is as well Bartók died before he could say the last word, leaving his life, like his work, an unsolved problem—a question and not an answer.[37]

[35] Henry Boys, 'Béla Bartók: 1881–1945', MT 86 (1945), 329–31.
[36] 'Béla Bartók: An Estimate and Appreciation', *Tempo*, no. 13 (Dec. 1945), 2–6.
[37] 'M. Béla Bartók', *The Times*, 15 Nov. 1945.

SELECT BIBLIOGRAPHY

Abraham, Gerald: 'Bartók and England', in Todd Crow, ed., *Bartók Studies* (Detroit, 1976), 159–66.

Anon.: 'Bartók Béla meleg ünneplése Londonban', *Prágai Magyar Hírlap* (Prague), 15 Oct. 1927.

Anon.: 'The Listener's Music: The Musical Capital of Europe', *Listener*, 2 Oct. 1935, p. 591.

Anon.: 'M. Béla Bartók: Hungarian Composer', *The Times* (London), 28 Sept. 1945.

BBC Handbook 1938 (London, 1938).

Bartók, Béla: *Hungarian Folk Music* (London, 1931).

Bartók, Béla, jun.: *Apám életének krónikája* (Budapest, 1981).

——, 'Bartók and the Visual Arts', *New Hungarian Quarterly*, no. 81 (Spring 1981), 44–9.

——, 'Béla Bartók's Diseases', *Studia Musicologica*, 23 (1981), 427–41.

——, *Bartók Béla műhelyében* (Budapest, 1982).

——, ed.: *Bartók Béla családi levelei* (Budapest, 1981).

Bator, Victor: *The Béla Bartók Archives: History and Catalogue* (New York, 1963).

Baughan, E. A.: 'A Hungarian Genius?', *Saturday Review*, 1 Apr. 1922, pp. 331–2.

Benkő, András: 'Romániában megjelent Bartók-interjúk', in Ferenc László, ed., *Bartók-dolgozatok 1981* (Bucharest, 1982), 272–361.

Berend, I. T., and Ránki, G.: *Hungary: A Century of Economic Development* (Newton Abbot, 1974).

Blom, Eric: 'Bartók, Béla', in H. C. Colles, ed., *Grove's Dictionary of Music and Musicians* (3rd edn. London, 1927), i. 232–5.

——, *Stepchildren of Music* (London, 1925).

Blythe, Ronald: *The Age of Illusion* (London, 1963; rpt. 1983).

Bónis, Ferenc, ed.: *Bartók Béla élete: Képekben és dokumentumokban* (Budapest, 1980).

Boult, Adrian C.: *My Own Trumpet* (London, 1973).

Boys, Henry: 'Béla Bartók: 1881–1945', *Musical Times*, 86 (1945), 329–31.

Breuer, János: 'Bartók and the Arts', *New Hungarian Quarterly*, no. 60 (Winter 1975), 117–24.

Brittain, Sir Harry: *The ABC of the B.B.C.* (London, n.d. [1932]).

Browne, Arthur G.: 'Béla Bartók', *Music and Letters*, 12 (1931), 35–45.

Bull, Eyvind H.: 'Pointed Paragraphs: The Music of Béla Bartók', *Music News* (Chicago), 7 June 1934.

Calvocoressi, M.-D.: 'Hungarian Music of To-day', *Monthly Musical Record*, 52 (1922), 30–2.

——, 'Béla Bartók: An Introduction', *Monthly Musical Record*, 52 (1922), 54–6.

——, 'Béla Bartók', *Musical News and Herald*, 11 Mar. 1922, p. 306.

——, 'Bartók's "Duke Blue Beard Castle" ', *Monthly Musical Record*, 54 (1924), 35–6.

——, *Musical Taste and how to Form it* (London, 1925).

——, 'Music of the Day: A Conversation with Bartók', *Daily Telegraph* (London), 9 Mar. 1929.

——, *Musicians Gallery* (London, 1933).

——, 'Bartók's Cantata Profana', *Listener*, 23 May 1934, p. 883.

[Capell, Richard] R. C.: 'Music Notes: Mr. Béla Bartók's Bombardment', *Daily Mail* (London), 3 Dec. 1923.

Capell, Richard: 'An Unavowed Romantic: Béla Bartók at the Queen's Hall—Barbaric Music', *Daily Telegraph* (London), 9 Nov. 1933.

——, 'Béla Bartók and Others', *Daily Telegraph* (London), 21 June 1938.

Chadwick, Richard: 'Alban Berg and the BBC', *British Library Journal*, 11 (1985), 46–59.

Chapman, Ernest: 'Béla Bartók: An Estimate and Appreciation', *Tempo*, no. 13 (Dec. 1945), 2–6.

[Colles, H. C.] H. C. C.: 'Béla Bartók: Sonata for Violin and Piano—A Foreign Language', *The Times* (London), 18 Mar. 1922.

Colles, H. C., and Cruft, John: *The Royal College of Music: A Centenary Record, 1883–1983* (London, 1982).

Copley, Ian A.: 'Warlock and Delius—A Catalogue', *Music and Letters*, 49 (1968), 213–18.

——, *The Music of Peter Warlock: A Critical Survey* (London, 1979).

Cowell, Henry: 'Bartók and his Violin Concerto', *Tempo*, no. 8 (Sept. 1944), 4–6.

Cox, David: *The Henry Wood Proms* (London, 1980).

Demény, János: 'Bartók Béla tanulóévei és romantikus korszaka', *Zenetudományi tanulmányok*, 2 (1954), 323–487.

——, 'Bartók Béla művészi kibontakozásának évei (I): Találkozás a népzenével (1906–1914)', *Zenetudományi tanulmányok*, 3 (1955), 286–459.

——, 'Bartók művészi kibontakozásának évei (II): Bartók Béla megjelenése az európai zeneéletben (1914–1926)', *Zenetudományi tanulmányok*, 7 (1959), 5–425.

——, 'Bartók Béla pályája delelőjén: Teremtő évek—világhódító alkotások (1927–1940)', *Zenetudományi tanulmányok*, 10 (1962), 189–727.

——, ed.: *Béla Bartók Letters* (London, 1971).

——, ed.: *Bartók Béla levelei* (Budapest, 1976).

——, ed.: 'Korrespondenz zwischen Bartók und der holländischen Konzertdirektion "Kossar" ', *Documenta Bartókiana*, 6 (1981), 153–229.

Dille, Denijs: 'Vier unbekannte Briefe von Béla Bartók', *Oesterreichische Musikzeitschrift*, 20 (1965), 449–60.

Dille, Denis: *Thematisches Verzeichnis der Jugendwerke Béla Bartóks, 1890–1904* (Budapest, 1974).

Evans, Edwin: 'Half-Time in England', *Modern Music*, 3. 4 (May 1926), 10–15.

——, 'Bartók, Béla', in Walter Willson Cobbett, ed., *Cobbett's Cyclopedic Survey of Chamber Music* (London, 1929), i. 60–5.

Fassett, Agatha: *Béla Bartók's Last Years: The Naked Face of Genius* (London, 1958).

Fenby, Eric: *Delius as I knew him* (London, 1936).

Földes, Andor: 'Rehearsing with Bartók', *New Hungarian Quarterly*, no. 81 (Spring 1981), 21–5.

Foss, Hubert J.: *Music in My Time* (London, 1933).

——, and Goodwin, Noel: *London Symphony* (London, 1954).

——, 'The Musical Press in England To-day', *Music and Letters*, 11 (1930), 128–40.

Foulds, John: *Music Today* (London, 1934).

Gillies, Malcolm: 'Bartók in Britain: 1922', *Music and Letters*, 63 (1982), 213–25.

——, 'Bartók, Heseltine and Gray: A Documentary Study', *Music Review*, 43 (1982), 177–91.

——, 'Grainger in London: A Performing History', *Musicology Australia*, 8 (1985), 14–23.

——, 'A Conversation with Bartók: 1929', *Musical Times*, 128 (1987), 555–9.

——, and Gombocz, Adrienne: 'The "Colinda" Fiasco: Bartók and Oxford University Press', *Music and Letters* (forthcoming).

Godfrey, Dan: *Memories and Music* (London, 1924).

Gombocz, Adrienne, and Somfai, László, ed.: 'Bartóks Briefe an Calvocoressi (1914–1930)', *Studia Musicologica*, 24 (1982), 199–231.

Gorham, Maurice: *Broadcasting and Television since 1900* (London, 1952).

Gray, Cecil: 'Béla Bartók', *Sackbut*, 1 (1920–1), 301–12.

——, 'Zoltán Kodály', *Musical Times*, 63 (1922), 312–15.

——, *A Survey of Contemporary Music* (London, 1924).

——, *Sibelius* (London, 1931).

——, *Peter Warlock: A Memoir of Philip Heseltine* (London, 1934).

——, *Predicaments* (London, 1936).

——, *Contingencies and Other Essays* (London, 1947).

——, *Musical Chairs* (London, 1948).

Hadow, Sir Henry, *et al.*: *New Ventures in Broadcasting* (London, 1928).

Hawkes, Ralph: 'Béla Bartók: A Recollection by his Publisher', in *Béla Bartók: A Memorial Review* (New York, 1950), 14–19.

Henry, Leigh: 'Liberations: Studies of Individuality in Contemporary Music, III: Béla Bartók and the Analysis of Racial Psychology', *Egoist*, 1 May 1914, pp. 167–9.

Heseltine, Philip: 'The Modern Spirit in Music', *Proceedings of the Musical Association*, 45 (1918–19), 113–30.

——, 'Modern Hungarian Composers', *Musical Times*, 63 (1922), 164–7.

——, *Frederick Delius* (London, 1923).

Hibberd, Stuart: *'This—is London'* (London, 1950).

[Holland, A. K.] A. K. H.: 'M. Béla Bartók', *Liverpool Daily Post*, 31 Mar. 1922.

Hull, A. Eaglefield, ed.: *A Dictionary of Modern Music and Musicians* (London, 1924).

——, *Music: Classical, Romantic and Modern* (London, 1927).

Juhász, Vilmos: *Bartók's Years in America* (Washington DC, 1981).

Kennedy Michael: *The Hallé Tradition* (Manchester, 1960).

Kentner, Louis: 'M. Béla Bartók', *The Times* (London), 15 Nov. 1945.

Kenyon, Nicholas: *The BBC Symphony Orchestra* (London, 1981).

Kroó, György: *A Guide to Bartók* (Budapest, 1974).

Lampert, Vera: 'Zeitgenössische Musik in Bartóks Notensammlung', *Documenta Bartókiana*, 5 (1977), 142–68.

Lukács-Popper, Mária: 'Bartók our House-guest', *New Hungarian Quarterly*, no. 84 (Winter 1981), 103–6.

Lutyens, Elisabeth: *A Goldfish Bowl* (London, 1972).

Lyle, Watson: 'Béla Bartók: A Personal Impression', *Musical News and Herald*, 19 May 1923, pp. 495–6.

Macleod, Joseph: *The Sisters d'Arányi* (London, 1969).

Maine, Basil: *The B.B.C. and its Audience* (London, 1939).

Matthay, Tobias: *The Act of Touch in all its Diversity* (London, 1903).

Nettel, Reginald: *The Orchestra in England: A Social History* (London, 1946).

Newman, Ernest: 'Musical Chart-Reading', *Sunday Times* (London), 7 Oct. 1945.

——, 'The Bartók Quartets', *Sunday Times* (London), 11 Nov. 1945.

Orga, Ateş: *The Proms* (Newton Abbot, 1974).

Parrott, Ian: 'Warlock in Wales', *Musical Times*, 105 (1964), 740–2.

Pollatsek, László: 'Béla Bartók and His Work', *Musical Times*, 7 . (1931), 411–13, 506–10, 600–2, 697–9.

Rees, C. B.: *One Hundred Years of the Hallé* (Norwich, 1957).

Reith, J. C. W.: *Broadcast Over Britain* (London, 1924).

——, *Into the Wind* (London, 1949).

Rubbra, Edmund: 'Béla Bartók's Second Piano Concerto', *Monthly Musical Record*, 63 (1933), 199–200.

Scholes, Percy: 'Is Bartók Mad—Or Are We?', *Radio Times*, 9 Dec. 1927, pp. 525–6.

——, 'The Music of Today', *Radio Times*, 18 May 1928, pp. 285–6.

Sessa, Anne Dzamba: *Richard Wagner and the English* (London, 1979).

Smith, Robert: 'Béla Bartók in Wales', *Welsh Music*, 4. 2 (Autumn 1972), 11–15.

[Somfai, László, ed.]: 'Vier Briefe Bartóks an Philip Heseltine', *Documenta Bartókiana*, 5 (1977), 139–41.

Stevens, Halsey: *The Life and Music of Béla Bartók* (New York, 1953; 2nd edn. New York, 1964).

Suchoff, Benjamin, ed.: *Béla Bartók Essays* (London, 1976).

Sutcliffe, Peter: *The Oxford University Press: An Informal History* (Oxford, 1978).

Szigeti, Joseph: *With Strings Attached: Reminiscences and Reflections* (New York, 1947).

——, 'Working with Bartók', *Music and Musicians*, 11. 8 (Apr. 1963), 8–9.

Taylor, Colin: 'Peter Warlock at Eton', *Composer*, no. 14 (Autumn 1964), 9–10.

Thompson, Kenneth: *A Dictionary of Twentieth-century Composers (1911–1971)* (London, 1973).

Tóth, Aladár: 'Bartók külföldi útja', *Nyugat*, 15 (1922), 830–3.

Valkó, Arisztid: 'Bartók Béla tervezett londoni hangversenyének levéltári háttere', *Magyar Zene*, 18 (1977), 99–105.

Van Dieren, Bernard: 'Musical Microtomy', *Monthly Musical Record*, 61 (1931), 330–3.

Volly, István: 'Bartókné Pásztory Ditta', *Életünk*, 21 (1984), 807–23.

Whitaker, Frank: 'A Visit to Béla Bartók', *Musical Times*, 67 (1926), 220–3.

——, 'The Most Original Mind in Modern Music', *Radio Times*, 26 Feb. 1932, p. 504.

Wood, Henry J.: *My Life of Music* (London, 1938).
Woodhouse, George: 'Béla Bartók', *Musical News and Herald*, 1 Dec. 1923, p. 474.
Young, Patricia, *et al.*: *The Story of the Proms* (London, n.d. [1957]).

INDEX